Agrarian Change
and
Small Farmers

About the Editors

Dr. Kaustubha Nand Bhatt is a researcher in Social Sciences with specialization in environmental economics and social development issues. Dr Bhatt has directed and authored/co-authored 21 research projects in inter-disciplinary issues. Most of these projects has been sponsored and funded by International Labour Organizations, the World Bank, UNDP, UNESCO, Planning Commission, U.P., Tata Chemicals, Sriram Foundation, National Literary Mission, Adult Education Directorates, U.P., MP, Maharashtra, State Primary Education Boards of U.P., Bihar and Hill Development Board, U.P. Dr. Bhatt has published five books and three dozen research papers in Journals and Edited Volumes. He has widely travelled around the globe on his academic sojourns to present research papers, deliver lectures and chair sessions in prominent universities of the world: University of Melbourne (2009), University of Western Australia, Perth (2009), Curtin University, Perth (2009), Limerick University, Ireland (2010), University of Roma Tre, Rome, Italy (2011), Civil Services Commission, Male, Maldives (2012), National Institute for Development Administration, Bangkok, Thailand (2012), and University of New Mexico Conference at Wisconsin, Madison, USA (2012), Bahrain Institute of Public Administration, Manama, Bahrain (2013).

Pradeep Bhargava is a professional researcher and has pursued research particularly in the areas of poverty, vulnerability, education, institutions and organizations. He has also studied civil society response to various problems. He worked at South Asia Studies Centre, University of Rajasthan, the Centre for Policy Research, New Delhi and the Institute of Development Studies, Jaipur. He was a Mid Career Fellow at the Woodrow Wilson School, Princeton University and a Visiting Fellow at the Institute of Development Studies, University of Sussex. He has authored and co-edited several books including Political Economy ot the Gulf States, New Economic Policy and the Vulnerable Sections, Policies for Agricultural Development and Poverty Reduction in Developing Countries : Experiences from Asia and Africa and has published in various journals. He engages in policy dialogue with Governments and interacts with Civil Society.

Agrarian Change and Small Farmers

Super Markets, Viability and Food Policy

Edited by

K. N. BHATT
PRADEEP BHARGAVA

Published for
G.B. PANT SOCIAL SCIENCE INSTITUTE,
ALLAHABAD
by
CONCEPT PUBLISHING COMPANY PVT. LTD.
NEW DELHI-110059

ISBN-13 : 978-93-5125-033-3

First Published 2014

Published and Printed by

Concept
Since 1974

Concept Publishing Company Pvt. Ltd.
A/15-16, Commercial Block, Mohan Garden
New Delhi-110059 (India)
T: 25351460, 25351794, *F*: 091-11-25357109
E: publishing@conceptpub.com, W: www.conceptpub.com
Editorial Office: H-13, Bali Nagar, New Delhi-110 015, India.

Cataloging in Publication Data--*Courtesy:* D.K. Agencies (P) Ltd. <docinfo@dkagencies.com>

Symposium on 'Agrarian Change and Small Farmers' (2012 : Govind Ballabh Pant Social Science Institute)
Agrarian change and small farmers : super markets, viability and food policy /edited by K.N. Bhatt, Pradeep Bhargava.
p. cm.
Includes bibliographical references and index.
ISBN 9789351250333

1. Farms, Small--India--Congresses. 2. Supermarkets--India--Congresses. 3. Sustainable agriculture--India--Congresses. 4. Food security--India--Congresses. 5. Nutrition policy--India--Congresses. 6. Agriculture and state--India--Congresses. I. Bhatt, K. N. II. Bhargava, Pradeep. III. Govind Ballabh Pant Social Science Institute. IV. Title.

DDC 338.10954 23

Preface

Recent years have witnessed depeasantization in India. Viability of small systems of farming has been questioned more in such regions where agriculture uses more of new seeds, chemical fertilizers and pesticides. Not as much where small farmers use little of these inputs and are more environment friendly. In the neo-liberal environment the existence of small farmer has been put to threat. It is uncertain as to how supermarkets will relate to the small farmer. Furthermore, there is a need to diversify the food basket and turn back to traditional organic farming using indigenous seeds. And in these circumstances our food policy needs to be rewritten.

This book is an outcome of a two days Symposium on 'Agrarian Change and Small Farmers' organized by the Govind Ballabh Pant Social Science Institute, Allahabad on March 11-12, 2012. We publish here thirteen revised papers authored by scholars from different disciplines. We would like to express our thanks to all contributors of papers and participants of the Symposium for their valuable contribution. We are thankful to the Indian Council of Social Science Research, New Delhi for partly funding the Symposium. Our thanks are due to Concept Publishers, New Delhi for bearing responsibility for publication of the book.

Allahabad
July 2013

K. N. Bhatt
Pradeep Bhargava

Contents

Preface v
List of Contributors ix
Acronyms xi

Section I
INTRODUCTION

1. Introduction 3
Pradeep Bhargava and *K. N. Bhatt*

Section II
SMALL FARMERS AND SUPER MARKETS

2. Do Asset Poor Farmers Gain from Market Reforms? 19
Nilabja Ghosh and *Shipra Pratap*

3. Small Farmers, Evolutionary Economics and the NAP: In Search of a Discrete Theoretical Basis 38
Pinaki Chakraborti and *Rik Chakraborti*

4. Adoption of Innovative Agricultural Practices across Supply Chain among Small-Scale Vegetable Growers 60
Jabir Ali and *Tribhuvan Nath*

5. Efficiency of Price Discovery in Agricultural Commodity Trading in India 75
Tanuj Nandan and *Niti Nandini Chatnani*

Section III
SMALL FARMERS AND VIABILITY

6. Strategy for Sustainable Development of Small Farmers 89
Amarendra Pratap Singh and *R. P. Singh*

7. Agrarian Change in Mountain Economies: The Constraints and Policy Options 112
I.C. Awasthi

8. Towards Improving Viability of Small Farmers 136
Brajesh Jha

9. Small Farms and Economic Viability in Uttar Pradesh 154
Vidyarthi Vikas

10. Agriculture in Odisha: Prospects for Growth and the Policy Imperatives 173
Prashanta Ch. Panda and *Sudhakar Panda*

Section IV
SMALL FARMERS AND FOOD POLICY

11. Nutritional Security through Diversified Food Production 187
Rita Singh Raghuvarshi

12. Climate Change and Sustainable Agriculture: Finger Millet for Food Security and Nutrition 201
K.N. Bhatt

13. Agrarian Society, Economic Reforms and the Concerns for Marginals: Opportunities and Challenges from Rural India 222
Ashish Saxena

14. Contemporary Debates on Food Security in South Asia 242
Sunil Kumar Tripathi

Section V
CONCLUSION

15. Epilogue: Alternative Ways for the Small Farm 259

Index 264

List of Contributors

Dr. Ashish Saxena, Department of Sociology, University of Allahabad, Allahabad-211 002, U.P. Email: ashish.ju@gmail.com

Dr. Brajesh Jha, Agricultural Economics Unit, Institute of Economic Growth, Delhi University Enclave, Delhi-110 007.
Email: brajcsh@icgindia.org

Dr. I. C. Awasthi, Institute of Applied Manpower Research, Planning Commission Government of India, Sector A-7, Plot No. 25, Institutional Area, Narela, New Delhi-110 040. Email: icawasthi@rediffmail.com, icawasthi@nic.in

Dr. Jabir Ali, Centre for Food and Agribusiness Management, Indian Institute of Management, Prabandh Nagar, Off Sitapur Road, Lucknow-226013, Uttar Pradesh. Email: jabirali@iiml.ac.in

Dr. K. N. Bhatt, G.B. Pant Social Science Institute, Jhusi, Allahabad-211019 U.P. Email: knbhatt1@rediffmail.com

Dr. Nilabja Ghosh, Agricultural Economics Unit, Institute of Economic Growth, Delhi University Enclave, Delhi 110 007.
Email: nila@iegindia.org

Professor Pinaki Chakraborti, Department of Economics, University of Burdwan, Golapbag, Burdwan-713 104 (West Bangal). Email: c.pinaki@gmail.com

Dr. Rik Chakraborti , Ph.D student, Department of Economics and Finance, The University of Wyoming, USA. Email: rchakrab@uwyo.edu

Professor Pradeeep Bhargava, Director, G. B. Pant Social Science Institute, Jhusi, Allahabad-211 019 (U.P.).
Email: pradeep@gbpssi.org.in

Prashanta Ch. Panda, Associate Professor of Economics, School of Liberal Studies, Deen Dayal Petroleum University, Gandhinagar, Gujarat. Email: pc_panda04@yahoo.co.in

S. Panda, Former Prof. of Economics, Utkal University, Bhubaneswar, Orissa. Email: pandasudhakar1945@rediffmail.com

Professor Rita Singh Raghuvanshi, Dean, College of Home Science, G.B. Pant University of Agriculture and Technology, Pantnagar, Uttarakhand. Email: deanhsc1@gmail.com

Professor R.P. Singh, Director, AMITY Business School, and Director QAE, Amity University, 5, Viraj Khand, Gomati Nagar, Lucknow. Email: rpsingh@lko.amity.edu

Sunil Kumar Tripathi, G.B Pant Social Science Institute, University of Allahabad, Jhusi, Allahabad-211 019.
Email: tripathisunil156@gmail.com

Dr. Amarendra Pratap Singh, Indira Gandhi National Tribal University, Amarkantak, M.P. Email: drapsinghigntu@gmail.com

Dr. Tanuj Nandan, Department of Management Studies, MLLNIT, Allahabad -211 001. Email: tanujnandan@gmail.com

Tribhuvan Nath, Centre for Food and Agribusiness Management, Indian Institute of Management, Prabandh Nagar, Off Sitapur Road, Lucknow-226013, Uttar Pradesh. Email: tribhwan@gmail.com

Dr. Niti Nandini Chatnani, Indian Institute of Foreign Trade (IIFT), New Delhi. Email: nitinandini@iift.ac.in

Vikas Vidyarthi, G.B. Pant Social Science Institute, Allahabad University, Jhusi, Allahabad - 211 019. Email: vidyarthiv@rediffmail.com

Dr. Vandana Shiva, A-60, Hauz Khas, New Delhi-110 016.
Email: vshiva@giasd101.vsnl.net.in

Acronyms

AAU	-	Assam Agricultural University
ADM	-	Archer Daniels Midland
AERCs	-	Agro-Economic Research Centres
ANFP	-	Adjusted Net Farmer Price
APMC	-	Agricultural Produce Market Committee
APMCA	-	APMC Acts
AU	-	Andhra University
CACP	-	Commission on Agricultural Costs and Prices
CAGR	-	Compound Annual Growth Rate
CBOT	-	Chicago Board of Trade
CCCC	-	Climate Change Compliant Crop
CMI	-	Corporate Marketing Intermediation
CSO	-	Central Statistical Organization
CSR	-	Corporate Social Responsibility
DFPCL	-	Deepak Fertilizers and Petro-Chemicals Limited
DM	-	Direct Marketing
ECA	-	Eastern and Central Africa
ECARSAM	-	Eastern and Central Africa Regional Sorghum and Millet-Network
EGPs	-	Employment Generation Programmes
FAO	-	Food and Agriculture Organization
GATT	-	General Agreement on Trade and Tariffs
GCARD	-	Global Conference on Agricultural Research for Development
GIPE	-	Gokhale Institute of Politics and Economic
GMO	-	Genetically Modified Organism
GRG	-	Growers' Representative Group
HPU	-	Himachal Pradesh University

HYV	-	High Yielding Variety
IADP	-	Intensive Agriculture District Programme
ICDS	-	Integrated Child Development Services
ICRISAT	-	International Crops Research Institute for the Semi-Arid Tropics
ICTs	-	Information and Communication Technologies
IDD	-	Iodine Deficiency Disorders
IFPRI	-	International Food Policy Research Institute
IMF	-	International Monetary Fund
IMR	-	Infant Mortality Rate
IPCC	-	Intergovernmental Panel on Climate Change
IPM	-	Integrated Pest Management
I-O	-	Input-Output
JNKV	-	Jawaharlal Nehru Krishi Vishwavidyalaya
KCC	-	Kisan Credit Cards
KVIC	-	Khadi and Village Industries Corporation
KVKs	-	Kisan Vikas Kendras
LBW	-	Low Birth Weight
LPG	-	Liberalization, Privatization and Globalization
MC	-	Marketing Cost
MDM	-	Mid-Day Meal
MCX	-	Multi-Commodity Exchange
MINEFF	-	Market Inefficiency
MNCs	-	Multi-National Corporations
MGNREGA	-	Mahatma Gandhi National Rural Employment Guarantee Act
MPCI	-	Monthly Per Capita Income
MRG	-	Margin
MSCALE	-	Marketing Scale Micro,
MSMED	-	Micro, Small and Medium Enterprises Development
NAFTA	-	North American Free Trade Agreement
NAP	-	New Agricultural Policy
NAPM	-	National Alliance of People's Movements
NARS	-	National Agricultural Research System
NCAER	-	National Council of Applied Economic Research
NCDEX	-	National Commodity & Derivatives Exchange
NCEUS	-	National Commission for Enterprises in the Unorganised Sector

NFHS	-	National Family Health Survey
NGOs	-	Non-Government Organizations
NMCE	-	National Multi Commodity Exchange
NRAA	-	National Rain-fed Areas Authority
NRI	-	National Research Institute
NSSO	-	National Sample Survey Organization
NFP	-	Net Farmer Price
NFR	-	Net Farm Returns
NSDP	-	Net State Domestic Product
OAEs	-	Own Account Enterprises
OBCs	-	Other Backward Castes
PCRM	-	Poisson Count Regression Model
PDS	-	Public Distribution System
PEM	-	Protein Energy Malnutrition
PMAG	-	Price Magnification
PRI	-	Panchayati Raj Institution
PSUs	-	Public Sector Undertakings
R&D	-	Research and Development
RB	-	Rythu Bazaar
REDS	-	Rural Economic Development Survey
RM	-	Regulated Markets
RNF	-	Rural Non-Farm
RNFEs	-	Rural Non-Farm Enterprises
RRBs	-	Regional Rural Banks
RUDA	-	Rural Non-Farm Development Agencies
SAT	-	Semi-Arid Tropics
SCs	-	Scheduled Castes
SEZ	-	Special Economic Zones
STs	-	Scheduled Tribes
SGDP	-	State Gross Domestic Product
SHGs	-	Self Help Groups
TFP	-	Total Farmer Price
TNC	-	Trans-National Corporations
TP	-	Terminal Price
TRIPs	-	Trade Related Aspects of Intellectual Property Rights
UNESCO	-	United Nations Educational, Scientific and Cultural Organization
UPA	-	United Progressive Alliance

US	-	United States
USDA	-	United States Department of Agriculture
WAST	-	Product Wastage
WHO	-	World Health Organization
WIR	-	Worldwatch Institute Report
WTO	-	World Trade Organization
ZCE	-	Zhengzhou Commodity Exchange

Section I
Introduction

1

Introduction

Pradeep Bhargava and *K. N. Bhatt*

Indian agriculture is a home for small farmers. They cultivate 44 per cent of operated land and constitute 85 per cent of total holdings as counted in the Agricultural Census 2008-09. Their numbers are on the rise as population increases holdings get fragmented: 70 per cent holdings were small covering 21 per cent operated land in the count of 1970-71. The opinion on whether small farmers are viable and could survive in the new liberal economic environment, which includes super markets for agricultural commodities, is divided. The home for small farmers has in recent years witnessed depeasantisation, which could be worrisome for some and inevitable for others. Small farmers are vulnerable given first of all their size, all the more, a reducing size.

Comparative advantage of small farmers over large commercial farms lies in low transaction costs of working with motivated family labour, low cost of supervising household labour, intensive local knowledge as well as saving on marketing due to self-provisioning. On the other hand large farmers have comparative advantage in lower costs of transaction with outside world in input, output and credit markets. Moreover, small farms may be a disadvantage if new technologies require more capital inputs and high level of education.

The viability of small farmers rests on productivity, and the income they derive from their lands, among other opportunities.

Farm Size and Productivity

The relationship between farm size and productivity could be inverse, positive or scale neutral. The implications of each are different. Those who empirically establish an inverse relationship argue that there should be more small farms in the future on grounds for efficiency and growth and hence advocate for policies that support land reforms and land ceilings. Assuming that technology is neutral to scale, policies are recommended for increasing access to resource base and support services to small farmers that substantially affect productivity of small farmers. It is argued that they could use available resources with them more intensively and judiciously. Another argument in their favour is that capital cost to the economy will be less if the choice is for small farms.

While not refuting the empirical evidence of higher productivity in small farms, some argue that this happens because low levels of technology are used by all farmers and with advances in technology the inverse relationship would disappear. The small farmers try and eke out a living from their tiny holdings and there is a desperate struggle for survival at below poverty level incomes. Thus, efficiency argument is disputed and it is argued instead that measures to create more small farms would deepen and perpetuate extreme levels of poverty.

On the other hand, some research reports point out that small scale farming to be inefficient in India by factoring in supervision costs, risks, credit market imperfections and scale-economies associated with mechanizations. Besides, in the long run the small farms may witness decline in productivity resulting from intensive cultivation and inability of the small farmer to replenish the soils due to lack of capability to invest, which is directly related to incomes of small farms.

That, the small farms have low (below poverty level) incomes has remained an established fact since the 1970s and in an established fact in the 2010s. And it has been suggested for the last four decades that such small farmers should have some non-farm incomes in order to survive. Thus, at the small farm level the productivity has increased and so has the cost of cultivation. The income gains of productivity do not offset the increase in costs as much that farms move beyond the poverty line. Thus, the hybrid seed, fertilizer, phytosanitary technology has at least not been able to raise farm incomes above poverty lines of the small farms. There are other reasons for low incomes as well like indebtedness.

Indebtedness of many small farmers is a result of price shocks or susceptibility to weather conditions: seasonal droughts and/or floods. One of the fallouts of indebtedness for a sustained period is expressed as a desire to exit agriculture, being forced to sell or lease out land or even commit suicide. In NSSO 2003 study 40 per cent farmers expressed their desire to leave farming. The process of depeasantisation in a state like

Punjab known to be as agriculturally rich began since early 1990s and gathered momentum since 2000. It is reported that more than 2 lakh small and marginal farmers have left farming in the State due to economic distress. These trends are important to track to understand the small farmers and what happens to them. First, the gains from Green Revolution started receding in the 1990s particularly in the crop sector. This period coincides with the period of liberalization wherein withdrawal of State support and decline in public investment in agricultural sector takes place. Second, structural problems continue to burden small farmers more than large farmers: access to institutional credit is limited and there remains high reliance on moneylenders; restricted reach to technology, extension services and skills; lesser gain from public irrigation and electricity; near absence of tenancy reforms, lesser bargaining power; skewed access of public goods such as public irrigation and electricity, markets and institutions. Third with agriculture becoming more market oriented since late eighties increasing imperfections and volatility in input and output markets has enhanced vulnerability of small farmers, affecting profitability.

The World Bank is generally of the view that the emerging new markets provide big opportunities for even the small farmers. This view is elaborated in development literature in a producer centric manner with the small farmer in the nucleus facing a number of challenges or barriers to modern markets: access to market information, bargaining power corresponding with low volumes of production, infrastructure in rural areas, financial, physical and human capital and low level of trust between users and buyers. This approach to development leads to conclusions such as proposing policies to the government to facilitate market access for small farmers , formation of farmers group, ensure infrastructure and provide essential services. The expected return is that the small famer would be expected to join the Super Market Chain.

At the same time there is increasing evidence that supermarket procurement models could have a negative impact on the small farmers even in the developed countries. This evidence is provided by investigative reports commissioned by EU Parliament and the US Department of Agriculture and International Lead Agencies. Super Markets are known to impose their supply chain requirements on to supplier and dictate terms and demand. In countries where Super Market concentration is high, a food producer may not be able to sell outside the Super Market value chain. Super Markets operate on a low profit margin, which would require keeping prices low for the consumer, increase quality of goods sold and push down costs. The Super Markets in turn pass these costs to the farmers, they increase the costs in risk of farming as they only deal with those farmers who are willing and able to carry these.

Small farmers would need strong linkages and reliable access to a range of resources and services is crucial if they are to survive in increasingly competitive globalized agri-food markets. Farmers organizations backed up by strong national policies may play a significant role in both service provision and advocacy.

The other side of the story is that almost all the small farmers in India practice multi-diversified farming and grow a number of crops even on small and fragmented plots. Many of them undertake crop and livestock farming in a symbiotic manner. But this kind of farming does not necessarily yield enough returns for the sustenance and upward movement of small farm families. Land can be successfully cultivated unless it provides adequate return. Therefore, sustainability of farming system requires that it is economically viable, technically feasible, environmentally sound, socially acceptable, administratively manageable and politically desirable.

The small farmers are also better stewards of natural resources which offer much more sustainability in the long run. They can also intensify biodiversity through conservation and rejuvenation resulting into higher biological productivity and stability and sustainability of agriculture.

Access to resource base and support services substantially affect the productivity of small farmers even though the technology is neutral to scale. There is enough scope for improvement of small farmers as they use the available resources with them more intensively and judiciously. Therefore, small farmers are not a threat for agricultural development but they are big strength for the overall socio-economic development of the country. The former Prime Minister of India, Choudhary Charan Singh clearly indicated that if a crowded, capital-scarce country like India has a choice between a single 100 acres farm and forty 2.5 acres farms, the capital cost to the national economy will be less if a country chooses the small farms.

The net profit per acre is claimed to be quite high on bio-diverse small ecological farms than large and industrial farms which are known to exploit and deplete natural resources, make indiscriminate use of chemical fertilizers more energy and heavy machinery and pollute the atmosphere and destabilize the climate. Therefore, small bio-diverse farms based on internal inputs are the only promise for increasing agricultural productivity while the industrial model of high external inputs imposed on small farmers might lead to increased indebtedness and suicides. These claims have to be empirically established and have not been so far because these experiments are sparse and have not been achieved on a large scale so far.

With an objective of analyzing these growing concerns, G. B. Pant Social Science Institute, Allahabad organized a two days National Symposium on 'Agrarian Change and Small Farmers during March 11-12, 2012. It was partly funded by the Indian Council of Social Science Research, New Delhi. The symposium was devoted to acquire broad understanding of the growing concerns on agrarian change and small farmers in India. Experts from multiple disciplines participated in the symposium and presented their research papers with subsequent in-depth discussions on each aspect of the subject. In all, 13 papers finally revised after the symposium has been received from the experts belonging to multi-disciplinary boundaries. These papers are accordingly categorized in the three sections apart from the conceptual issues raised in the introductory chapter and concluding remarks subsequently at the end. The three major sub-themes are: (1) Small Farmers and Super Markets; (2) Small Farmers and Viability; (3) Small Farmers and Food Security.

In the last decade or so in some regions markets have started engaging small farmers through various interventions including rural service hubs, parts of value chain or contract farming. The government has suggested Group Approaches including companies. Banks are also considering Self Help Groups (SHGs) among other options. Super Markets expect strict standards, quality, timelines of supply, ability to adjust to changed demand, which small farmers mostly uneducated may find difficult to meet. Several crucial questions need exploration and policy initiatives. Are small farmers of India sufficiently capitalized and educated to meet the expectations of Super Markets and hence sustain development? Or as International Food Policy Research Institute (IFPRI) suggests that policy makers need to consider whether there are social reasons to support small farmers? Or should policy agenda involve social safety needs and facilitate good exits from farming for small farmers?

The second crucial question is the traditional small farmers, who grow mainly for subsistence, perhaps environment friendly, uses organic manure, own seeds still there? Are these possibilities for its existence? Is it possible that control over land, water and agricultural bio-diversity lies with decentralized agricultural communities who are able to control their livelihood, food and their natural resources? Finally how does agricultural policy relate to ecological security, food security and agricultural security to the small farmers? What could the traditional small farmer and modern small farmer learn from one another? Are small farmers prepared for Super Markets? Can Small farmers have food sovereignty? Are small farmers food secure? The questions involve several concerns like small farmer and agrarian structure, poverty, depeasantisation, crop mix and farm size productivity of small farmers, small hubs, value chain, credit, producer Associations, control of agricultural community over land, water and agricultural bio-diversity, etc.

Small Farmers and Super Markets

Introducing the theme, Nilabja Ghosh and Shipra Pratap observe that reforms in agriculture are intended to make the markets more efficient. The amendment in the Agricultural Produce Market Committee (APMC) Acts as suggested by the Central Government allows transactions to take place outside the regulated market premises and permits private companies to enter into the field of agri-marketing. In States that have initiated the reforms, several channels that permit direct marketing by farmers, corporate intermediation in market, contract farming and linkages with organized retailers are already emerging. Worldwide, a major criticism against marketing innovations relates to the restrictive nature of the new chains that constricts entry.

In India, any lack of inclusiveness could be a serious weakness leading to enhanced inequality and precluding asset poor farmers especially the large majority of small farmers from claiming any benefit from reforms. By analyzing primary data on two emerging channels, Direct Marketing and Corporate Market Intermediation, collected from different parts of India, the study by Ghosh and Pratap finds that new channels are more efficient than traditional ones but while the first model is fairly inclusive, the second presents mixed performance and needs more attention.

The chapter by Pinaki Chakraborti and Rik Chakraborti examines the problem of exclusion of the small Indian farmers, from the standpoint of Evolutionary Economics and explores the New Agricultural Policy of India. Building up a discrete mathematics model using weighted digraphs and the shortest path algorithm, of a two sector multiple product economy, the chapter shows that exclusion of small farmers results from a two-layered 'selection' process. On one layer some aspects of economies of scale favours the big farms and firms. Small agents are generally outcompeted resulting in exclusion of the small.

However, Chakraborti and Chakraborti argue that this is not really a 'natural selection' process. It is rather activated through an artificial drive towards becoming big. Now, there is a coevolving layer of selection process that disfavours the 'small' further by the selection of inputs not quite conducive to smallness, such as the cultured seeds. Not being from the natural routes, their management needs adverse fertilizers, water absorption, pesticides and so on. The process leads to environmental degradation going against the small farms more than the big farmers. The study finally suggest for significant amendments based on the evolutionary envisioning of Indian agriculture with reference to sufferings of the small farmers, in the New Agricultural Policy (NAP) for the United Progressive Alliance (UPA) Government of India.

Jabir Ali and Tribhuvan Nath have noted that small-scale farmers are considered as primary stakeholders in agriculture in developing countries. Their farm practices and technology adoption determines the success of agricultural supply chain. High-value vegetables are perishable in nature and it is necessary that greater coordination be obtained in how food is produced, processed, marketed, and consumed. Vegetable cultivation practices of the small holders are market driven, therefore, they tend to adopt innovative farming practices that lead to high farm earnings. Despite various initiatives to improve production and productivity in vegetables, fundamental problems and concerns related to high post-harvest losses, which require a better approach to link on-farm activities to processing, marketing, and distribution.

Ali and Nath analyzed the adoption behaviour of small-scale vegetable growers on innovative farm practices by across supply chain, cultivation, post-harvest practices and marketing of vegetables based on a primary survey of 556 vegetable growers in six districts of Uttar Pradesh. Findings suggest that the small holders are inclined to adopt innovative farm practices including modern farm inputs, however, they continue to face complex challenges in its access due to socio-economic, farm level and institutional constraints. The regression analysis was also carried out to identify the most likely factors affecting adoption of innovative farm practices across the chain. The results have important implications for public and private agencies including non-government organizations (NGOs) to ensure availability, access and successful adoption of innovative agricultural practices to the small holders for the supply chain efficiency and effectiveness.

Tanuj Nandan and Niti Nandini Chatnani profess that the introduction of futures trading in the agricultural commodity segment in India has been the subject of much debate relating to the impact it exerts on the pricing of these commodities. It has been argued that futures trading lend excessive volatility to the prices of agricultural commodities, thereby increasing risk across the value-chain in these commodities. At the same time, it is also counter-argued that futures trading tend in agricultural commodities to stabilize prices, by creating a broad-based market. In the Indian context, the introduction of futures trading has resulted in an efficient price-discovery process. The introduction of exchange-based electronic trading platforms has facilitated communication of price-related information to traders across the world. At the same time, the ability of entering into trades and offsetting them without the physical interchange of commodities has brought in price-risk hedgers and arbitrageurs, who work towards minimization of price differentials, and thereby lead to price stability.

Further, Nandan and Chatnani have clearly noted that this market allows the participation of a large number of traders with market intelligence and price judgment, which ultimately results in efficient price discovery. India's agricultural sector has been characterized by low levels of informational and pricing efficiency, as is evidenced by the fact that only 25-30 per cent of the price paid by the consumer for agricultural produce is actually realized by the farmer, whereas in developed markets like the US, the farmer's share can be 60-65 per cent. In India, the benefits offered by exchange-based commodity trading towards the enhancement of these profit margins can be extended to large and small farmers alike, by assigning greater importance to the role of aggregators and marketing federations, while reducing dependence on other intermediaries.

Small Farmers and Viability

Amarendra Pratap Singh and R.P. Singh describe the contribution of Indian agriculture in national GDP during last 60 years which has constantly declined from 55 per cent in the early 1950s to nearly 15 per cent in 2010 but its contribution to total employment during the same period has declined marginally from 70 per cent to 50 per cent. Indian agriculture is dominated by marginal and small farmers accounting for more than 83 per cent of total holdings and 42 per cent of the cultivated area. Moreover, average size of holding has also declined from 2.28 to 1.21 hectares during 1970-2005. Further, the number of marginal and smaller holdings is likely to increase further due to subdivision of families but will continue to play a very important role in ensuring the food and nutritional security to the ever increasing population. Following a series of neoliberal economic reforms in 1991, India's GDP has doubled, but despite this, 53 million more people now go to bed hungry every night. Farming is becoming less profitable and a large number of farmers are not interested in farming. The food and nutritional security of the country very much depend upon these small farmers. But, they face several physical, institutional, infrastructural, technological constraints.

Singh and Singh further highlight the intense debate on the inverse relationship between farm size and productivity which over time changed due to differences in agro-climatic conditions, level of infrastructure development, adoption of superior modern technologies and access to various input support services. Hence, there is a need to make them sustainable through region-specific farmer-centric agricultural policy to make them economically viable. The sustainability of these small holdings require necessary infrastructure, effective land reform policies such as removing restrictions of tenancy, redistribution of quality land to needy farmers, reconsolidation of tiny land holding in a separate block with full support for irrigation, improved seeds, fertilizer, credit at affordable cost, and more importantly strong network of extension. Besides, formation of Farmers Interest Groups, Farmers Clubs, need to be encouraged. However, focus should be not only diversification of cropping system but more on farming system. Farmers should not be only spectator but an active partner in designing and implementation of programmes. Moreover, agriculture policies should focus on more drought stress and low input based cropping system, which are also highly nutritious. Also, equal partnership can be developed with private sector to increase the access to market, technology, and global opportunities. Besides, efforts need to be made to create more employment opportunities outside agriculture to divert the surplus manpower from agriculture sector.

I.C. Awasthi analyzes the mountain economies which are noted to have increasingly confronted with the challenges and problems concerning to sustainable livelihoods, mountain agriculture, persistent poverty, fragile environment, and female drudgery. Traditionally, hill people have been meeting their food and other non-food requirements through crop cultivation, livestock rearing and collection of forest produce. The cultivable land being scarce getting further fragmented with the increase of population and it has hard pressed the people to look for alternative livelihood strategies. The alternative strategies consisted of extension of cultivation on hitherto uncultivated marginal land (though limited), improvement in quality of livestock, diversification of agriculture into high value crops such as horticulture and non-food crops for the market.

But these coping strategies have limitations to provide food security to the people on a sustainable basis due to low productivity of resources and the problems associated with the hill specificities on the one hand and increasing population on the other.

Awasthi further advocates that the manufacturing base in any case is miniscule in these areas; no large industries can be set up in the hilly region due to several constraints. Village and household industries based on livestock, food processing and natural resource have been promoted but it could not generate self-reinforcing and self-sustaining forces due to small scale operation, little quantity, dispersed production and lack of access to appropriate institutions. Moreover, these industries have been subjected to increasing competition from the similar products from the plain areas that are superior in terms of quality and price much cheaper than the local products. This has had an adverse impact on the local economy. When agriculture and industrial base is weak this would obviously have far reaching implications to the hill economies. The development strategies, therefore, requires to be distinct in different hill economies in view of regional specificities that need to be focused and promoted on mountain resources that has clear comparative advantages (hydro power, tourism, horticulture, and other niche products).

Brajesh Jha in his chapter on 'Towards Improving Viability of Small Farmers' says that in India over 84 per cent of agricultural holdings are less than 2 hectares. Bulk of these holdings is not viable on their own; farmers operating such small holdings have to attend many non-agricultural activities for their livelihood. The modern technologies in agriculture however require more intense involvement of farmers in farming. In dearth of sufficient off farm income opportunities farmers at times migrate to other regions leaving their land under certain sub-optimal arrangements. The small and marginal farmers together now accounts for more than 40 per cent of cropped area of agriculture in the country; productivity on these farms is, therefore, important not only for viability of these farms but also for the growth of agriculture in the country. In any strategy to improve viability of such holdings growth in rural non-farm sector is important. Analysis of secondary information from different sets of data suggests that for growth of productive employment in rural non farm sector, manufacturing is important. In the recent years though manufacturing is being emphasized, importance of manufacturing in rural vicinity is not being recognized. Such initiatives are however important to improve viability of small holding agriculture and also for the balanced development of the country.

Vidyarthi Vikas examines the Small Farms and their economic viability in Uttar Pradesh and argues that Productivity is bound to grow slower in agriculture than non-agriculture sector because of greater impediments to specialization and the division of labour in agriculture production. Now this situation has emerged in Indian agrarian structure due to itself and tiny holding. Average farm size of holding gradually decreased. The increasing trend of small farm is due to several factors. This trend of persistence create situation more difficult and raise a question regarding economic viability of small holding and subsistence of small farmer. While the concept of 'economic holding' is a dynamic phenomenon, a small holding can become a viable economic holding and there is still room for a static definition of economic size at any point of time and places. But most of the farmers grow some definite crops in Uttar Pradesh as a whole. Therefore, there is an opportunity to identify any particular size of land holding. Based on the systematic analysis of NSS data and field work, Vidyarthi suggests that there are considerable opportunities to enhance farm-productivity and to make small farms economically viable.

Prashanta Ch. Panda and Sudhakar Panda suggest that notwithstanding the declining contribution of agriculture to Gross State Domestic Product in Orissa, the major defining trend of Orissa's economy will be agriculture for a number of reasons including the assurance of food security to the people. Its development will impact state's efforts to create sources of livelihood and eradicate rural poverty. Given the agro-climatic zones and rising demand for varieties of fruits and vegetables in the State for domestic consumption and sale in other States, diversification of agricultural production holds great prospects for the farming community including the small and marginal farmers who constitute around 85 per cent of total farm households. The conditions that matter for the development of agriculture will be not only improved infrastructure, but also the timely supply of improved seeds, credit, water, chemical fertilizers, extension services and above all a good market that offers the producers a remunerative price. Along with it, efforts have to be made to offer the farmers insurance cover against crop loss and compensation for destruction of standing crops by the wild animals. All this calls for a coordinated approach to development of agriculture by all the stake holders.

Small Farmers and Food Policy

Green revolution has transferred the country from food deficient state to food sufficient nation argues Rita Singh Raghuvanshi. She informs that agriculture currently accounts for 14 per cent in National GDP, having declined from 24 per cent, which is a worrying sign. In spite of large production, malnutrition is rampant in the country. It is a cause of concern, sufficient food production and availability has not percolated to individual households, therefore, it has resulted in food insecurity and poor nutritional status. Challenges are to increase diversified food production for nutrition security. This can be done by giving due importance to the production of course grains, i.e., amaranth, buckwheat, millets, horse gram, rice bean and maize, etc., in addition to local fruits and vegetables; roots and tubers; local herbs and spices and other underutilized and wild foods. These products have supported life of people over centuries, they have varied nutrients and phytochemicals in different amounts available in different seasons thus supported health and nutrition of the population. Agriculture scientists should pay attention for large cultivation of these items to meet the population demand of food for nutrition security through women and small farmers.

K.N. Bhatt examines inter-linkages between incoming climate change and Sustainable Agriculture. He analyzes *Ragi* (Finger Millet) as climate change compliant crop with ecological, food, biodiversity, nutrition, health security benefits. Feasibility of finger millet production, product diversification, markets, and its impact on the aspects of social and distributive justice for poor and small farmers are explored. Finally, the paper searches for institutional alliances and seeks answers for how finger millet production can be initiated in rural India by various stakeholders, including public sector enterprises.

Bhatt argues that all millets including the wonder crop finger millet are no doubt bio-diverse highly nutritive and climate change compliant crops. Finger millet may provide great socio-ecological bonus to farmers: climate change bonus, bio-diversity bonus, water conservation bonus, organic food bonus. As the food crop is known for its high nutritive values with major and micro nutrients, it may be introduced in public distribution system throughout the country. Similarly, finger millet foods may be introduced in Indian defence forces, railways, integrated child development services schools, colleges and hostels, baby foods, public sector undertakings. Finger millet is grown almost in all parts of India and people have skills for its production. It is indeed an equity ensuring crop system. Finger millet is also known for its good quality livestock feed for enhanced milk production as its stems contain maximum quantity of lactose. People centred and people directed Research and Development activities on finger millet will go a long way in the development of this valuable food crop with its numerous environmental benefits. PSUs may initiate Corporate Social Responsibility activities for the development of finger millet as a sustainable food crop of the future.

Ashish Saxena established that Indian society is highly diversified and based on hierarchical principle resulting into unequal citizens, regional inequality and rural-urban divide. Globalization usually benefits people who can play the game, but the resourceless, uneducated, and those making a living through traditional production activities, come under various kinds of pressures. It is thus clear that the economic reforms of the Indian economy has dramatically influenced social life in the country—marginalizing countryside vis-à-vis urban; peasantry vis-à-vis farmers in the countryside. The state has mounted policies of economic liberalization, claiming that the policies will ultimately benefit the entire population but marginalized groups remain adversely affected with uneven regional development, wage disparities and increased level of inequality. Even the large sections of the backward classes and castes remain extremely hard-hit by inflation, contradiction in secure rural employment, the virtual impossibility of entry into lucrative export markets, continual poor levels of literacy, health and living standards, and overall social and political marginality.

In the present era of privatization, further highlights Saxena, the state's indirect withdrawal from 'welfare' of service sector people, particularly *dalits,* became vulnerable. The shift of decision making power into the hands of trans-national institutions like the WTO, International Monetary Fund, World Bank and others has severely reduced the sovereignty of national government and resulted in a very serious drift. We find that the 'marginals' are at the lowest end of the economic and social ladder and are quite obviously at the receiving end of the elitist culture promoted by urban world and by globalization. The discourse in this chapter thus revolves around—implications under globalization for the rural India. To put it differently, it reflects the level of inequitable inclusion of the rural population and the marginals. Gradual decline in the nation-state policy to tackle the "social concern" (welfare goals) for marginal peasantry is emphasized along with an impression that the new regime promotes more inequality, starvation, indebtedness, suicides, deprivation and depeasantisation vis-à-vis agriculture and consolidation of their holdings in favour of corporate farming.

Sunil Kumar Tripathi says that the food security is a major issue for any country as its scarcity can cause considerable distortions. Achieving food security has been a national goal for all the countries in the world. In Asia addressing the high levels of poverty, food insecurity, and child malnutrition is an urgent goal from an ethical point of view. Achieving food security for all depends on policy action and institutions that address the causes of food insecurity, malnutrition and unsustainable management of natural resources. The specific policies that will be most appropriate will vary according to the local and national circumstances. This chapter does not seek to be a comprehensive literature review. Instead, key co-relates underlying food insecurity, including declining agricultural productivity, changing economic policies and challenges of climate change as seen in the literature are critically analyzed along the line of availability, access, utilization and stability. The study deals with the issues, options and challenges that the South Asian nations are facing in reference to food security with special focus on India and begins with introduction followed by outlining the contemporary understanding of food security situation in South Asia and the national context and priorities followed by discussion on issues of debate on food security in South Asia.

Section II
Small Farmers and Super Markets

2

Do Asset Poor Farmers Gain from Market Reforms?

Nilabja Ghosh and *Shipra Pratap*

Emerging Concerns in Post-Reform Markets

Reforms in agriculture intend to make markets for agro-commodities more efficient than they are. Though the prevailing regulatory framework was intended to ensure fairness, in reality the rigid structure, entry barriers and an overarching role of state supervision only helped to tilt the balance of market power between the producers and the traders in favour of the latter. Worse, the system led to a proliferation of intermediaries operating inefficiently and even redundantly in the marketing channels. The relaxations implicit in the amendments to the APMC Acts (APMCA) as suggested by the Central government were meant to encourage interested new parties to enter the market as buyers and to allow them greater leeway in conducting their transactions such as to purchase outside the premises of the Regulated Markets, strike prior contracts with producers over prices and qualities and transact with producers directly as final users of products. Most importantly, all the buyers have the opportunity to 'contest' one another[1] to ensure that prices are determined by demand and supply forces of the market.

Agricultural marketing, however, far from being a simple economic issue, is not only complex but is deeply political in implication. The concept of efficiency, intensely associated with reforms, and its measurability have been questioned as orthodox and a constructs of contemporary institutions (Hariss-White, 1996). The replacement of an existing structure by an unknown and evolving one, therefore, is a fertile ground for skepticism. Agriculture being a State subject under the Indian Constitution, reforms repeatedly encountered impediments arising from electoral compulsions of State governments. Moreover, agricultural reforms worldwide have not been free from misgivings either. In the wake of trade liberalization, evidences of intermediary margins, perceived partly as entrepreneurs' 'reputational' rents and partly as returns to managerial skills, getting inflated and translating to enlarged ratio of the earnings of non-production workers to those of production workers in agriculture has been instrumental in inspiring market reforms (Goldberge and Pavnic, 2007, Bardhan *et al.*, 2009).

All this indicates an inherent inefficiency of the prevalent systems and the intense deprivation of the actual producers from the market derived gains. Yet assessments on reforms in developing countries have shown ambivalent results (Negassa and Jayne, 1997, Wu and McErlean, 2003). The apprehensions surrounding market power, entry barriers and transparency issues have not spared the developed countries too (Treleavan, 1975; McCalla and Schmitz, 1979; Yong and Hobbs, 2000, 2002; Hamilton, 1997). Agro-ecologists have argued that intermediation in emerging channels can mean shifts in farming practices in search of profit that undermines soil quality (Rosset, 2011). This chapter is designed to address some of these concerns from evidences emerging from India.

The recent spurt of reforms in many States in India and the consequent emergence of new marketing channels raise several concerns such as: (i) whether the emergent markets are a step towards greater competition or a mere invitation to powerful players in an unequal market; (ii) whether they will benefit only exclusive sections and exaggerate rural inequality; (iii) whether the new system, by creating a threat of competition for the buyers, offer viable options to the producer-seller; and (iv) if the threat created by each competitive agent on another has made a difference to trading practices. This chapter addresses these concerns by examining the structure and operation of two select emerging marketing channels in the post-reform situation and the tendency of the asset poor farmers to participate in these markets. Although a number of such channels have emerged in various parts of the country such as contract farming, retail chain, corporate marketing and direct marketing this study could consider two channels namely Direct marketing (DM) and Corporate Marketing Intermediation (CMI). The study is based on primary data supplemented by on-site observations and perceptions expressed by players in the questionnaire based surveys and in focus group discussions.

Evolution and Reforms of Markets: A Background

A large volume of goods traded world-wide are handled in the internal trades of countries. Early studies undertaken in the middle of the 20th century (Yang, 2000, McBryde, 1947, Berry, 1967 and Hodder, 1965), recognized that the market place was a fundamental focal point of economic and social life in a peasant society. The importance of trading intermediaries increased as economic functions became more specialized relative to primitive person-to-person trading transactions in the underdeveloped agriculture and in conditions perceived to be ideal (Tax, 1953) and close to perfect competition but owing to shortage of personal capital, the individual trader is likely to be merely a link in a long chain with other intermediaries (Bromely, 1971). Such long marketing chains are today frequently condemned as inefficient with large elements of middle-man margins and marketing costs intervening between the producer and consumer prices. Markets that are formally free are known to be influenced by the distribution of economic power (Weber, 1978) in which small and asset poor farmers are like to be severely disadvantaged.

Even though small farm dominated subsistence farming was the norm, marketing of surplus grains was a traditional practice in India. Regulation of markets in India initiated by the Royal Commission on Agriculture in 1928 aimed to reduce exploitation of small and ignorant farmers but nevertheless, the markets remained uneven when India became independent. The marketing policy of independent India's government was meant to facilitate price determination by market forces rather than market power and to protect small farmers from exploitation. Although normative price decision was not the objective of regulation, the Government did intervene actively in the case of a few foodgrain staples to prevent prices from going down below a minimum threshold. Despite all the overt intentions, the system created vested interests, collusion with the regulators, barriers to entry, restrictions on producers in the sale of their own products, rigid specification of the place of sale and a regime of low prices that favoured the urban consumers and deprived producers. In the process, the food market and agriculture also remained starved of investment in modern technology and painfully slow in procuring information on market conditions. Failures of the system became manifest in various evaluations (*see* Acharya, 1997, 2004) which showed that traders succeeded in bidding for low prices when farmers hardly acted in collectives. Licensing requirements prevented many interested buyer from entering the market. Despite all the limitations, the farmer was compelled to bring the produce to the regulated markets (RM), travelling long distances and face poor facilities that resulted in spoilage of products and inconvenience. On the other hand, processors, exporters, retailers and even consumers found it nearly impossible to buy directly from producers.

While the Indian agricultural marketing system was constricted in a narrow frame of wholesale-retail and middlemen dominated network with extensive State regulations, marketing in many western countries especially the US evolved into complex and highly sophisticated structures marked by substantial managerial and technological innovations. The requirement of a vibrant and dynamic marketing structure to meet the challenges emerging out of globalization in the post WTO period came out in the report of an Inter-Ministerial Committee[2] leading to the formulation of a Model APMCA. Finalized in 2003 and circulated to enable States government to amend their own APMCA, the model Act was meant to allow more flexibility for farmers to sell their produce, permit private companies to intermediate in the market and improve the efficiency of the market functioning. However, several States had reservations on accepting all or parts of the Model and till date not all States have reformed agricultural marketing by passing the new Act.

Data and Methodology

Primary information is collected by the Agro-economic Research Centres (AERC) located in the Himachal Pradesh University (HPU) at Shimla in Himachal Pradesh, Gokhale Institute of Politics and Economic (Deemed to be University) at Pune (GIPE) in Maharashtra, Jawaharlal Nehru Krishi Vishwavidyalaya (JNKV) at Jabalpur in Madhya Pradesh, Assam Agricultural University (AAU) at Jorhat in Assam and the Andhra University (AU) at Visakhapatnam in Andhra Pradesh. The first three AERCs studied the CMI model in the emerging channels category while the last two, i.e., those in Assam and Andhra University addressed DM model, that was found to be gaining popularity in the States. All the five States studied have amended their APMCA.

It was planned that each centre would study two crops of horticultural nature by drawing random samples of participants. Fruits and vegetables are potential commercial crops in which the private players are expected to operate intensely in particular. Lists of participants were prepared based on information from the corporate agent and local authorities and the samples were drawn at random giving representation to different holding sizes. Random samples of farmers operating in a common traditional channel in the same area were also drawn for comparison. All this was supplemented by focus group discussions with market authorities, traders and consumers and available secondary data. The reference period is 2009-10.

Table 2.1: Sample Details

State	*District*	*Block*	*Products*	*Sample size EMC (TMC)*
Direct Marketing Channel				
Assam	Tinsukhia	Hapjan and Kakapathar	Orange	50 (50)
Andhra Pradesh	Visakhapatnam	NA	Banana, Brinjal	30 (30) 30 (30)
Corporate marketing intermediation Channel				
Himachal Pradesh	Shimla	Rohru	Apple	50 (50)
Maharashtra	Pune	Satana	Pomegranate, Onion	5 (35) 12 (35)
Madhya Pradesh	Sehore	Sehore	Soyabean	37 (35)

Note: Figures in parentheses are the respective indicators in the sample of traditional channel participants.

Since the emerging market models are only at an inception stage, the sampling design was strongly restrained by the actual presence of the relevant emerging channel in the area. Selling by farmers to a company named Deepak Fertilizers and Petro-Chemicals Limited (DFPCL) is considered as participation in the relevant CMI channel for both the crops in Maharashtra. Since the channel is yet in its infancy, its acceptance being extremely limited, the Centre could locate only 17 farmers who disposed their products in this channel, these products being Onion of which Maharashtra is a major producer and Pomegranate known for the demand for its tasty juice and health benefits. The AERC in Madhya Pradesh could not identify two crops and in fact not even one horticultural crop that has found its place in any emerging channel. Only one crop, i.e., Soyabean, which is an oilseed crop rapidly gaining importance in the country and also acclaimed for positive health effects, could be studied and the much discussed. ITC-e-Choupal's intermediation is the emerging channel that is studied. In Himachal Pradesh, Apple, the main commercial fruit crop in the State, is selected for study. The sample is drawn from the Rohru Block in Shimla which has the maximum area under apple in the State and the marketing channel is intermediated by Adani Agrifresh Limited, a private company. Under the DM model, a farmers' market called *Rythu Bazaar* (RB), which is a chain that has gained popularity in southern India, is studied for two horticultural crops Banana and Brinjal in Andhra Pradesh while the AAU found the presence of farmers in remote areas of district Tinsukia in Assam, selling oranges individually and in groups directly to processors and consumers. Farmers from Hapjan and Kakapathar Blocks in Tinsukhia districts were surveyed. These details are provided in Table 2.1.

It was also found that channels tended to be scale sensitive so that it was not possible to ensure representation of all farm sizes in studying any channel. Besides the average land holding size also varied across regions in India, which made uniform categorization difficult. Thus the sample size under each conventional class[3] was restrained by the maximum availability of farms in that class participating in the channel. Efficiency is generally assessed by comparing the price the farmer receives or the so called 'farmer price' with what is deemed to be the final or the terminal price, which is most frequently the consumer price. While the farmer price and cost are calculated as the sample average in all cases, choice of the terminal price can be made based on the most common destination of the product. In Maharashtra the company buys from the farmer only a part of the product and the terminal user is the price at which the product is sold to the mall collected from DFPCL. The corresponding traditional channel price is the consumer price collected at a retail market. In Himachal Pradesh the marketing channel starts with the purchase of Apples through local collection centres of the company but the target destination is much more indeterminate. Since much of the procurements made by Adani actually reaches the traditional channel at some intermediate point[4], wholesalers in Delhi's Azaadpur market is considered as the terminal user of apples for both channels.

In Madhya Pradesh the producer disposes of Soyabean either through the traditional channels or the e-choupal and it is Soyabean oil that the ultimate consumer buys. Purchase of soyabean grains by consumer is rare in the surveyed area. The processor is, therefore, treated as the terminal purchaser of Soyabean produced by farmers. The RB is the obvious answer in Andhra Pradesh where the consumer buys directly in the emerging channel and in the traditional channel, the terminal prices are collected from a regular urban retail market called *Poorna* market located in the same city of Visakhapatnam. The AAU in Assam has considered three variant emerging channels: (i) where farmers sell to consumers individually, (ii) Farmers route their products to consumers *via* Growers' Representative Group (GRG), and (iii) where the GRG sells to processing units that further sell processed products to consumers. The corresponding channels in the traditional route are producer to consumer *via* middlemen through three different chains. The statistics reported in this study are simple averages in the two channels. The terminal price is the retail price collected from RM Committee.

The methodology consists of comparing sample averages in the two marketing channels for assessing efficiency and the representation of small farmers in the samples and ownership of key economic, informational and social assets. Primarily, the assessment is built on certain quantified indicators for measuring the efficiency and the inclusiveness of the channel.

TFP (Total Farmer Price) = Price received by the farmer.
NFP (Net Farmer Price) = TFP—marketing cost borne by farmer.
ANFP (Adjusted Net Farmer Price) = Net Price received by farmer adjusted for rejection and losses.
MSCALE (Marketing Scale) = Value of product offered for sale in the specified market.
WAST (Product Wastage) = Amount product loss on farm, in transit, storage and at retail.
MC (Marketing Cost) = Sum of marketing cost borne by producers and all intermediaries involved.
MRG (Margin) = Sum of margins earned by market intermediaries.
TP (Terminal Price) = Price paid by consumer or processor treated as final users.
NFR (Net Farm Returns) = Returns earned by producer per hectare of crop cultivated.
PMAG (Price Magnification) = TP/ANFP.
MINEFF (Market Inefficiency) = (MC+MRG)/ANFP.

The process of agro-marketing is embedded in agrarian relations and is deeply associated with rural class structures. Production too is intimately related to marketing so that integration of the entire process is widely admitted. Thus these markets are not seen in isolation in literature, rather they inspired studies that encompassed product markets, production, and other derived markets (such as labour, finance and land) grounded in institutional and Marxist approaches (Bhaduri, 1983, Bharadwaj, 1985). In general ownership of land and other assets is treated as a key indicator of rural class. While much of the literature is contextual to pre-capitalist agrarian economy and few studies of markets also addressed channels in the market especially in assessing the public sector marketing institutions, it is but natural that market liberalization would revitalize the focus on class impacts in context of emerging marketing channels in the liberalizing market. The relevance of the subject finds greater appeal in the backdrop of contemporary significance attached to participation and inclusiveness.

Emerging and Traditional Marketing Channels

In this section an overview is provided on how the emerging channels selected for the study function along with a description of the chains. This is set against the background of the traditional marketing set-up and the channels chosen for sampling in this category. Sample indications as seen in Table 2.2 clearly suggest that the emerging channels are more efficient than the traditional ones and DM appears to be more efficient than CMI even though individual performances vary. Price magnification (PMAG) is also higher under the traditional channel signifying that the share of producers in the consumer price is more under the emerging channel. Producers in the CMI can claim 63 per cent of the consumer rupee on the average whereas their share is 79 per cent in DM. The corresponding shares in traditional channels are 55 per cent in CMI and 52 per cent in DM. The corresponding shares in traditional channel are 55 per cent and 52 per cent in DM. That the farmers gain more from the emerging channels than the traditional ones is also partly borne out in Table 2.3 which shows that returns per hectare are higher among sample farms in the emerging channel on the average, and particularly in the CMI. The popular apprehensions about soil degradation also find no support if use of organic nutrients in soil is any indication. Belying our expectation that the DM patronized mostly by small farmers would draw small lots of products, the scale of marketing, i.e., volume offered for sale is larger in the emerging channel than in the traditional one by a factor of 1.8 times. This factor is only moderately higher in the CMI channel at 2.1. It is noted that producer till now do not concentrate all their marketable products in the CMI channel. In Maharashtra there is substantial diversification among alternative channels (Table 2.2). While producers often have suspicion about trader's bids, the corporate sector also does not enjoy their confidence.

Table 2.2: Performance of the Emerging Marketing Channels for Agricultural Products

State	*Corp*	*Marketing inefficiency*	*Price magnification*	*Marketing scale*	*Wastage*
Corporate Marketing Intermediation (EMC)					
Himachal Pradesh	Apple	0.46 (1.6)	1.46 (1.6)	619.83 (703.37)	2.6 (3.4)
Madhya Pradesh	Soyabean	0.18 (0.29)	1.18 (1.29)	33.64 (8.84)	3.6 (3.0)
Maharashtra	Onion	1.31 (1.26)	2.3 (2.26)	1187.67 (693.3)	10 (20.9)
	Pomegranate	0.39 (1.15)	1.39 (2.14)	3257.91 (1029.82)	9.7 (19.9)
Average		0.59 (1.08)	1.58 (1.82)	1274 (608.83)	6.5 (11.8)
Direct Marketing					
Andhra Pradesh	Banana	0.13 (0.97)	1.13 (1.75)	2374.9 (1009.3)	0.2 (0.8)
	Brinjal	0.03 (1.14)	1.336 (1.84)	726.2 (571.6)	5.9 (11.0)
Assam	Orange	0.336 (1.14)	1.35 (2.14)	85.5 (71.03)	0.7 (0.7)
Average		0.17 (1.08)	1.27 (1.91)	1062.2 (580.65)	2.27 (4.17)
Unit	Name	Ratio	Ratio	Rs. 000	% of product

Note: Figures in parentheses are the respective indicators in the sample of traditional channel participants.

Table 2.3: Gains to Farmers and Agriculture

Particulars	*Direct Marketing*			*Corporate marketing intermediation(EMC)*			
	Andhra Pradesh		*Assam*	*Himachal Pradesh*	*M.P.*	*Maharashtra*	
	Banana	*Brinjal*	*Orange*	*Apple*	*Soyaben*	*Onion*	*Pomegranate*
Returns from Land	108.1 (85.8)	42 (48.8)	1.3 (1.1)	4.4 (3.7)	1.7 (1.4)	1.8 (1.4)	11.5 (2.7)
Organic fertilizer use	545.8 (1308.2)	29633 (22212)	129.6 (117.5)	12325 (6260.0)	113.0 (70.0)	14061 (12567)	32857 (17326)

Note: Organic fertilizer use in Rs./hectare for Corporate Market Intermediation and in Kg./hectare for Direct Marketing samples. Figures in parentheses are the respective indicators in the sample of traditional channel participants.

The Traditional Market Channels as Perceived

We mentioned that regular marketing system had been languishing in different ways prior to the amendment of the APMCA. While regulation was malfunctioning in Andhra Pradesh, lack of regulation was the key shortcoming in Assam. In Himachal Pradesh the regulated markets under the traditional system was perceived as unfair especially to small farmers. In Madhya Pradesh and Maharashtra several weaknesses marked the RM related to infrastructural inadequacy but there were few misgivings about traders' practices in the traditional channels as against the CMI channels. Interestingly, upgradation of the regulated markets in the wake of competition was reported in these States. Small traders continue to be important sources of valuable information about market prices, participants in both markets in all the States studied being beneficiaries. Despite intentions, the public information system under AGMARKNET has not served any of the respondents. Traders are still dominant players especially in the case of onions in Maharashtra where the product has a wide market, transcending the country's border. Also, some of the drawbacks that hit the traditional channels relate to infrastructure like roads and utilities like reliable weighing facilities that the State only can provide and, therefore, afflict participants in both channels.

Direct Selling

Direct selling in the DM model is essentially based on producer to consumer sales, a practice that had prevailed in primitive marketing systems in developing economies (Epstein, 1982) before the appearance of specialization of functions, a hallmark of modern sophisticated economies. Yet this model finds a place in the emerging scenario because of its simplicity. This is an option available to producers who wish to bypass middlemen and thus avoid the ordeals of bargaining and exploitation resulting from asymmetric information and to avail the full amount of prices that consumers are willing to pay. The appeal of this market arises from a notion that a section of urban consumers may be willing to pay a premium for products that are brought fresh from the farms. It is also felt that emerging market scenario is likely to favour bulk sellers who attract processors and corporate buyers allured by economies of scale. DM would, therefore, be particularly attractive for small producers.

The sample States Andhra Pradesh (AP) and Assam present contrasting cases. AP has had a vigorously regulated marketing system that had become stifling and misdirected over time. The State has played an active role in promoting DM by setting up Rythu Bazaars where farmers, enrolled as members from surrounding villages, sold to urban consumers. On the other hand despite initiatives in Assam, market regulation did not really take off in this remote State lying on the foot of the Himalayas in the north-east. Sales of agricultural goods till now take place mostly in local markets and in road-side stalls. It is the recent amendment that seems to have mobilized regulations in the State. While local markets continue to be major site of sale in both channels, paradoxically the RM accounts for 75 per cent of the sales under the sample DM model. A second major area of difference between the two cases is the presence of group marketing and its rationale in Assam. Orange producers in Tinsukhia have always found it difficult to market their products from the remote growing areas and depended helplessly on traders who leased in their land or operated as wholesalers in urban markets. DM gives an opportunity to market products concertedly through self-help groups or GRGs. The collectivity is seen to empower producers in the market even to sell to processors who generally deal with bulk suppliers. Unlike in AP, where farmers selling individually in the RB bore all the costs but received the same price as the consumers paid (Table 2.4), in Assam, marketing cost was shared by a group of farmers and a small margin was reserved to cover administrative expenses to be incurred by the GRG. Large producers are reported to be relatively less interested in selling directly.

Table 2.4: Prices Declared in Specified Markets

Particulars	*Direct Marketing*			*Corporate Marketing Intermediation*			
	Andhra Pradesh		*Assam*	*Himachal Pradesh*	*Madhya Pradesh*	*Maharashtra*	
Rs/Quintal	*Banana*	*Brinjal*	*Orange*	*Apple*	*Soyabean*	*Onion*	*Pomegranate*
Total Farmer Price	426 (299)	1320 (1498)	2000 (935)	4428 (6833)	1910 (1850)	694 (711)	6100 (3819)
Consumer/ Terminal Price	426 (390)	1320 (2479)	2000 (2000)	6172 (8486)	2242 (2352)	1600 (1438)	8500 (7500)

Note: Figures in parentheses are the respective indicators in the sample of traditional channel participants

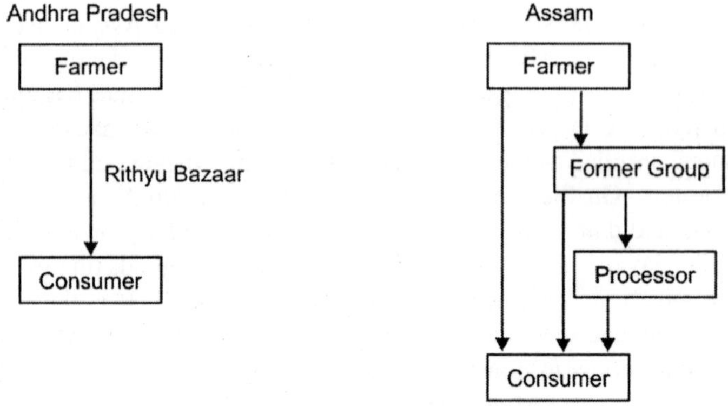

Fig. 2.1(a): Direct Marketing and Regulated Marketing

Corporate Market Intermediation

The three channels studied in this category in three different States are mutually different in their role and history but nevertheless share commonality. A key binding element is the profit motivation of an extraneous agency that has a capital base far stronger than individual traders can ever be associated with. The resource power also makes it possible for the companies to look for innovative and optimum ways of operation, invest in facilities and infrastructure that are cost effective and above all employ and develop a fleet of trained personnel to conduct the business. In all cases the corporate intermediary has pre-accumulated experience and expertise to work in agricultural markets.

DFPCL the CMI channel studied in Maharashtra was initially specialized in manufacturing fertilizers but later it diversified into agricultural extension and finally entered into marketing of farm products through its Agri-service division known as *'Saarrthie'*. A farmer can be enrolled as a member in *Saarthie* with a valid photo identity card on payment of a lifetime membership fee. The *Saarthie* provides marketing links with terminal buyers located both in India and abroad, helps farmers to acquire compliance with international market standards necessary to compete in the export market and generates resilience to price volatilities through its port storage infrastructure. It has linkages with processing industries and also has its own specialized processing facilities. While facilitation of exports is a central domain, the company also helps farmers in marketing farm produce within the country. The DFPCL procures fruits and vegetables, packs them and transports them in refrigerated trucks and sells them to exporters and organized retailers operating in Malls and Super Markets. The DFPCL does not have its own retail outlets although it has future plans in that direction. Its customers include Aditya Birla Retail Ltd, Metro Cash and Carry, TESCO India, Future Value retail Ltd. and exporters. The marketing channels provided by DFPCL are as follows:

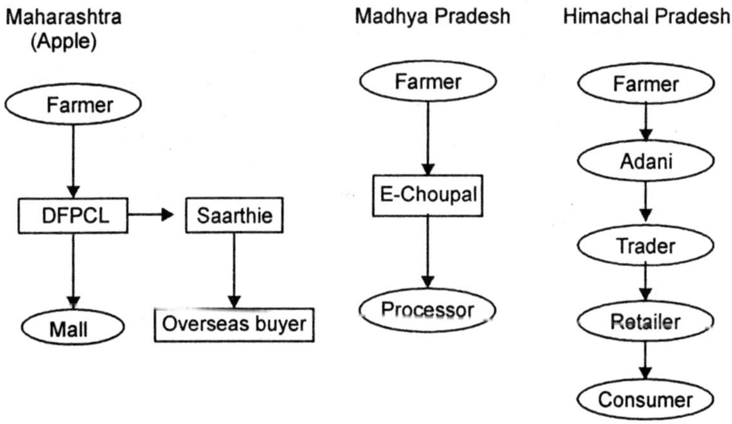

Fig. 2.1(b): Marketing Channels

Apple in Himachal Pradesh has traditionally passed through a number of hands before reaching the consumer, a fleet of traders hailing from Delhi (Azaadpur) markets being the major intermediaries. Recently, some private and public enterprises[5] have started providing organized marketing services. Adani Enterprises, a large Indian business group with diverse interests, is engaged in providing services to apple producers in Himachal Pradesh, where under the registered name of 'Adani Agrifresh Limited', it set up integrated storage, handling and transportation facilities geared for fresh produce. The group's managers have interacted with producers across the State and signed agreement with thousands of farmers for direct procurement.

The Adani group enrolls certain agents in the apple growing areas, who in turn enroll members among apple producers who would be willing to sell the produce to the group. The members are supplied with plastic crates free of cost for collection of apples that are brought to Adani stores. Since Adani sells a large part of the procured products to trader, till now the two channels are not sharply demarcated and the emerging marketing model also has to accommodate conventional intermediaries like *'mashakhor'* and the retailer who dominate the traditional channels. Adani too does not sell apples in the retail market.

The e-*choupal* initiative of the large company ITC Ltd., earlier a tobacco giant but today a highly diversified company, is well documented and is a popular case study in management courses. Although farmers are free to sell to the best buyer through the e-choupal, the company also purchases products from farmers. It is also launching a chain of rural malls and Sehore already has one on Indore-Bhopal highway. The choupal has good infrastructure in the form of market yard, canteen, parking space and drinking water supply while computerized weighing facilities are available to the sellers along with ATM banking.

Do Emerging Channels Accommodate the Asset Poor Producers?

Although the emerging channels appear to be more efficient than traditional ones a key question that begs answer for justifying their presence is whether market reforms is a step towards greater rural inequality and deprivation of the asset poor among the farmers in the post-liberalization India. Early signs of this predicament would be evident if the emerging channels are noted to exclude the relatively less endowed among the farmers in their set of participants. This section examines the extent of participation of the asset poor in the channels in these sets. The term asset-poor is defined by using a few select indicators but the share of small and marginal farmers is, as conventional in literature, accorded primacy. This is because these sections of farmers constitute over 80 per cent the farm holdings and their exclusion would hardly mean that a dent has been made by the reforms either in agriculture's developmental potential or in rural poverty. These farmers owing to paucity of landed assets have poor access to finance necessary for productive agriculture and are highly vulnerable to defaults and debt-traps. Figures 2.2(a) and (b) show that the average farm size of the sample farmer is more under the CMI channel cases in Maharashtra and Madhya Pradesh than the DM cases in Andhra Pradesh. However same indicator suggests that CMI has a small farm in its net while the average farm size is much larger in Assam than in Andhra Pradesh under DM. Also it is found in Figure 2.2(b) that the average farm size is determined by regional factors apart from channel choice since the averages under the traditional channels too follow the similar pattern.

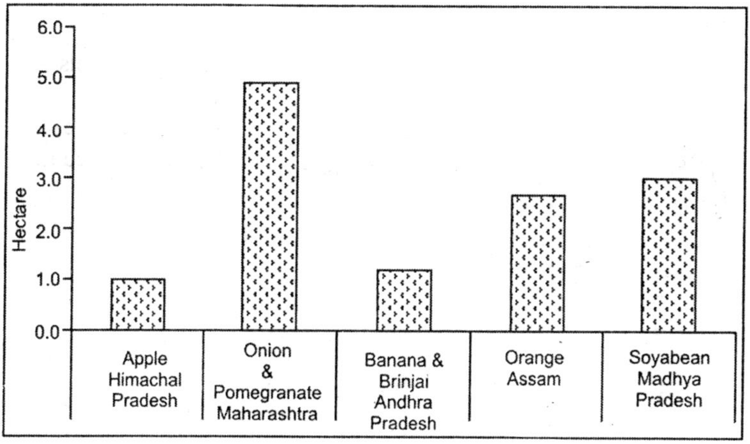

Fig. 2.2(a): Average Farm Sizes in the Emerging Market Channels

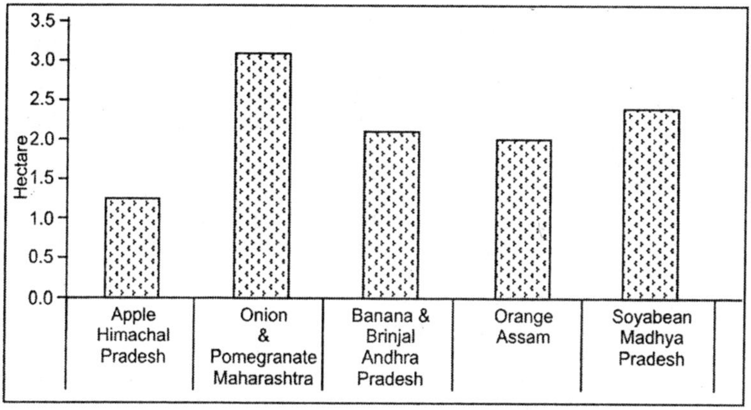

Fig. 2.2(b): Average Farm Sizes in the Traditional Market Channels

However land, despite being a prime source of economic power in rural India is admittedly not the only resource base. Although the poorer among the less land-endowed is more likely to be less endowed in other economic assets as well, any among the small farmers may be relatively rich in other assets while access to land need not necessarily translate to economic strength when productivity of the land determined by intrinsic properties of the land. Access to investible resources, cost of cultivation and managerial resource base has a strong effect on the economic outcome. Table 2.4, therefore, provides the share of backward classes (SC/ST) in the participant pool in the sample to represent social exclusion, the share owing farm asset (pump set) to capture agricultural asset base, the share of farmers having a vehicle (at least a two-wheeler) to indicate both affluence and mobility and the share of farmers owning a communication asset (mobile phone). The corresponding statistics for sample farmers in the traditional channel is given for comparison. The available data too suggests that the DM channel is more inclusive of the asset poor while the CMI except in Himachal has targeted the asset rich more dominantly. However the performance of Adani, a CMI which is efficient as well as inclusive, is notable.

Table 2.5(a): Corporate Marketing Intermediaries

State	%Backward classes	% Small and Marginal	% owning farm asset	% owning a 2 or 4 wheeler or both	% owning Mobile
Himachal Pradesh	24 (4)	88 (88)	—	12 (6)	100 (72)
Madhya Pradesh	5.4 (14.3)	8.1 (45.7)	97.3 (94.3)	81.1 (68.6)	78.4 (63.89)
Maharashtra	0 (5.7)	25.4 (42.9)	100 (100)	100 (77)	96.6 (94.3)
Direct Marketing	—	—	—	—	—
Andhra Pradesh	0 (10)	80 (53)	48 (97)	14 (?3)	100 (0)
Assam	0 (0)	44 (98)	20 (26)	30 (28)	92 (92)

Note: Figures in parentheses are the respective indicators in the sample of traditional channel participants.

Table 2.5(b): Comparison between Direct Marketing and Corporate Marketing Intermediation

Particulars	Direct Marketing			Corporate Marketing Intermediation			
	Andhra Pradesh		Assam	Himachal Pradesh	Madhya Pradesh	Maharashtra	
	Banana	Brinjal	Orange	Apple	Soyaben	Onion	Pomegranate
Returns from Land	108.1 (85.8)	42 (48.8)	1.3 (1.1)	4.4 (3.7)	1.7 (1.4)	1.8 (1.4)	11.5 (2.7)
Organic fertilizer use	545.8 (1308.2)	29633 (22212)	129.6 (117.5)	12325 (6260.0)	113.0 (70.0)	14061 (12567)	32857 (17326)

Note: Organic fertilizer use in Rs/hectare for Corporate Market Intermediation and in Kg/hectare for Direct Marketing samples. Figures in parentheses are the respective indicators in the sample of traditional channel participants.

The comparison between the two emerging channels namely DM and CMI and intra-model comparison demonstrate the diversity of implications that is inherent in reforms (Table 2.5).

Conclusions

Reforms in agricultural marketing are associated with the emergence of new marketing channels. Reforms can be successful and promising in India only if the emerging channels are inclusive enough to draw the participation of the asset poor farmers, especially the small farmers. Primary data collected in different parts of the country to study the performance of Direct marketing and Corporate Market Intermediation suggest that while both channels are efficient relative to traditional channels, inclusiveness of the channels is not assured. While the Direct Marketing Channel is more inclusive of the asset poor, corporate intermediation has produced mixed results. The samples in Maharashtra and Madhya Pradesh suggest that the participants in the emerging channels are privileged. Although the result is opposite in case of Himachal Pradesh, this could possibly be associated with local pressures exerted on the corporate intermediary to refrain from excessive selectivity that their demanding quality standards would indicate.

NOTES

1. An idea of a 'contestable' market was first proposed by Baumol, 1982.
2. An Expert Committee on "Strengthening and Developing Agricultural Marketing" under the chairmanship of Shri Shankerlal Guru was appointed by Government in December 2000.
3. Conventional farm size classes are broadly as follows: Marginal, operating up to one hectare of land, Small operating between one and two hectares, Medium operating two to ten hectares and Large who hold over ten hectares of land.
4. It may be noted that Adani's business has encountered local pressures and are yet evolving in the region. The company preferred to buy only the best quality apple grown mostly in upper altitudes but as farmers have resisted the continued dependence on traders for disposing the remaining produce and traders have also incited the process with their market power over farmers, Adani has accepted the practice of buying apples of all grades from producers and keeping the best for onward marketing, it sells the balance to traders at various points.

5. Container Corporation of India, Dev Bhoomi Cool Chamber Ltd., Indian Railway Board is some of the other operators in the State.

REFERENCES

Acharya, S. S. and N. L. Agrawal (2004). *Agricultural Marketing in India*, Oxford India, Delhi.

Acharya, S. S. (1997). "Agricultural Price Policy and Development: Some Facts and Emerging Issues", *Indian Journal of Agricultural Economics*, 52, 1, Jan-March.

Bardhan, Pranab, Dilip Mookherjee and T. Masatoshi (2009). "Middlemen Margins and Globalization, Paper presented at Seminar organized by *IFPRI*, mimeo.

Baumol, W.J. (1982). "Contestable Markets: An Uprising in the Theory of Industry Structure", *American Economic Review*, 72 (72), pp. 1-15.

Berry's, B. J. L. (1967). *Market Center and Retail Distribution*, Prentice-hall, Englewood Cliffs, New Jersey.

Bhaduri A., (1983). *The Economics of Backward Agriculture,* Academic Press, New York.

Bharadwaj, K. (1985). "A View on Commercialization in Indian Agriculture and the Development of the Capitalism", *Journal of Peasant Studies,* 12(1), pp. 7-25.

Bromley, R.J. (1971). "Markets in Developing Countries: A Review", *Geography*, 56(2), April: pp. 124-132.

Epstein Scarlett T. (1982), *Urban Food Marketing and Third World Rural Development*, Croom Helm, London.

Goldberg Pinelopi and Nina Pavcnik (2007). "Distributional Effects of Globalization in Developing Countries", *Journal of Economic Literature*, 65, March, pp. 39-82.

Harriss-White, B. (1996). "A Political Economy of Agricultural Markets in South India", Sage Publication, New Delhi.

Hodder, B.W. (1965). "*The Distribution of Markets in Yorubaland"*, Scottish Geographical Magazine, 81, p. 57.

McBryde, F. W. (1947). "Cultural and Historical Geography of Southwest Guatemala", Smithsonian Institution, *Institute of Social Anthropology*, Publication No. 4, Washington, D.C.

McCalla, A. F. and Andrew Schmitz (1979). "Grain Marketing Systems: The Case of the United States versus Canada," *American Journal of Agricultural Economics,* 61 (2), May, pp. 199-212.

Negassa, Asfaw and T. S. Jayne (1997). "The Responses of Ethiopian Grain

Markets to Liberalization", Working Paper 6, Ministry of Economic Development and Cooperation, Addis Ababa, January.

Rosset, Peter Michael, Braulio Machin Sosa, Adilen Maria Roque Jaime and Dana Racio Avila Lozano (2011). "The Campesino-to-Camesino agro-ecology movement of ANAP in Cuba: Social process methodology in the construction of sustainable peasant agriculture and food sovereignty", *Journal of Peasant Studies*, Online Publication, January13, *http://www.informaworld.com/smpp/title-content=t713673200*

Tax, S. Penny (1953). "*Capitalism: A Guatemalan Indian Economy*", Smithsonian Institution, Institute of Social Anthropology, Publication No. 16, Washington, D.C, pp. 15-18.

Treleavan, D.H. (1975). *Speech to Fifth International Grains Industry Course*, Winnipeg, can.: Canadian International Grains Institute, 10 Sept.

Weber, M. (1978). *Economy and Society: An Outline of Interpretive Sociology*, University of California, Berkley.

Wu, Ziping and Seamus Mc Erlean (2003). "Market Efficiency in Reformed Chinese Grain Marketing System", *China Economic Review*, 14, North Holland.

Yang, C.K. (1944). *A North China Local Market Economy: A Summary of a Periodic Markets in Chowping*, Hsien, Shantung, Institute of Pacific Relations, New York, Mimeo, p. 2.

Young, L.M. and J.E. Hobbs (2002). "Vertical Linkages in Agri-Food Supply Chains: Changing Roles for Producers, Commodity Groups, and Government Policy", *Review of Agricultural Economics*, 24, pp. 428-441.

3

Small Farmers, Evolutionary Economics and the NAP: In Search of a Discrete Theoretical Basis

Pinaki Chakraborti and *Rik Chakraborti*

I. Introduction

The small farmers in India have been under serious survival pressure for some time. The Ministry of Agriculture commissioned Situation Assessment Survey of the farming community, carried out by the National Sample Survey Organization. It brings out the distressed condition of India's small and marginal farmers in terms of income, expenditure or indebtedness (Narayanmurthy, 2006).

As Kumar (2006) noted, there is growing realization that the current cropping pattern and the strategy of increasing production is progressively becoming costlier in terms of rising costs of production, overexploitation and degradation of land and water resources; and in general, resulting in strain on the environment. It is being argued that growth in productivity of important crops has either declined or stagnated, whereas input intensity has been increasing over the years leading to declining total factor productivity. As farming in general has become costlier, the process discriminated against small farmers. Even though small and marginal Indian farmers were never in a very happy situation since Independence, their plight has increased manifold since the Green Revolution matured, and even more so since the New Economic Policy under globalization was launched. Kumar also observed that the last three decades of commercialization have heavily burdened the farmers with the use of certain inputs such as tractors, pump-sets and chemical fertilizers. As a result, per hectare return on land has started either stagnating or declining. In the given circumstances, small and marginal farmers have started leasing out land in favour of medium and large farmers, a phenomenon widely quoted in the literature as "reverse tenancy". Such a situation means that small farmers have been losing their land also to the big farmers for some time. Given the conventional attitude of farmers to land one easily understands that this can neither be desired consequences of policy authorities nor can be a result of optimization from free choice.

All this resulted from the economic policies to improve the conditions of Indian agriculture subsequently. We present a selected part of New Agricultural Policy (2011), NAP henceforth below.

The National Policy on Agriculture seeks to actualize the vast untapped growth potential of Indian agriculture, strengthen rural infrastructure to support faster agricultural development, promote value addition, accelerate the growth of agro business, create employment in rural areas, secure a fair standard of living for the farmers and agricultural workers and their families, discourage migration to urban areas and face the challenges arising out of economic liberalization and globalization. For this, among other things the following were suggested:

1. Greater private sector participation through contract farming.
2. Dismantling of restrictions on movement of agricultural commodities throughout the country.
3. Rational utilisation of country's water resources for optimum use of irrigation potential.
4. Capital inflow and assured markets for crop production.
5. Exemption from payment of capital gains tax on compulsory acquisition of agricultural land.
6. Adequate and timely supply of quality inputs to farmers.
7. High priority to rural electrification.
8. Setting up of agro-processing units and creation of off-farm employment in rural areas.

The policy is to promote technically sound, economically viable, environmentally non-degrading, and socially acceptable use of country's natural resources—land, water and genetic endowment to promote sustainable development of agriculture. It is clear that policies listed by 1 through 8 above, all favour, and/or oriented towards, facilitating large holdings and farmers. Looking at a deeper level through these eight points, one discerns three essential features. First, the threat emerging for small and marginal farmers, has been anticipated by the policymakers themselves, and hence comes the specific significance of point 8. The NAP is likely to inflict lack of livelihood for those to be excluded because of the policies, and so arises the need for new off-farm employments. Secondly, free and smooth functioning of national and international market forces is being ensured (points 2, 3, 4, 5). This would, in turn, imply entry of capital, in all possibilities foreign also, which under the current technological state, would favour large scale production, using cheap labour drawn from the 'reserve army' created by small farmers, losing their land to owners of big capital. Third, costlier availability of inputs are apprehended (policy 8).

It can be argued that this set of policies will aggravate further the small farmers' plight that they are already in. The distress has been in the literature for some time recently (Vyas, 2004; Wilson, 2002). If one looks at the efficacies of promulgated policies *ex post*, in Indian agriculture, as we noted earlier, one necessarily is baffled by inability to reach targeted outcome if not sheer failure ultimately, often within a short period of their operative regime. Question naturally arises why the policies generally miss the targeted section of the population. In the context of rampant suicide of small farmers in India recently, the question emerges to be a very significant one. In this chapter, it is intended to figure out the lacuna and look for a solution accordingly. Failure is not a novel experience in our history of policy outcomes. Moreover, this is not a unique experience in India only. Quite a few poor and developing countries have experienced similar adverse circumstances consequent upon positive policy interventions. (Please see Jacquet *et al.*, 2011).

Policies are devised by the national/State governments, who depend on their think tanks, research and advisory wings for suggestions. Such suggestions are generally based on research following the neoclassical tax-transfer economics of policy. Such exercises conventionally are empirical tests of hypotheses derived from application of comparative statics (occasionally dynamics) methods such as Envelope Theorem on relevant parameters for the question(s) examined. Thus policies are usually based on neoclassical optimization postulates. Also, any major policy change may be viewed as an innovation in the political economy of the imposing country, some change in the institutional structure of the associated economic space. Thus policies are signs of adaptation to the current pressures on survival of the economy and its people, may be a specific subset of them or all in general. Be that as it is, policy seems to be a certain selection node when the decision takers are faced with a multiplicity of possible scenarios with different degrees of pressures and net benefits of paths to be chosen by a policy mix selection which in turn, cannot be independent of policies adopted earlier.

This chapter intends to highlight this aspect with reference to the emerging adverse conditions of the small and marginal farmers in India. The main point that it makes is that the aforesaid theoretical basis of our policies is inappropriate; and hence, it is imperative now to premise on a realistic alternative theoretical-analytical structure, namely Evolutionary Economics, so that the resulting policy suggestions turn out to be really useful.

For an alternative approach this chapter appeals to Evolutionary Economics which does not require any smooth optimization over continuous, Inada-robust objective functions. The key concepts of this branch are used to understand the evolution of Indian agriculture and for analyses an alternative mathematical structure is suggested which does not require the continuity or convexity assumptions for existence of solutions. Such a structure is adapted from discrete mathematics. Graph theory is taken up for the purpose at hand. Evolutionary economics rules out any optimization and replaces that by 'satisficing' under bounded rationality, a premise much more relevant for real economies, especially as they are perennially under stress of some kind or other.

In the next section, the essential tenets of evolutionary economics are described. In section three the policy history of Indian agriculture is reviewed briefly with a view to understand the experiences of the small and marginal farmers in India. The fourth section narrates the correspondences of this history with the key concepts of Evolutionary Economics. Here it is argued that a more realistic analytical basis for a comprehensive policy framework capable of closely anticipating the appropriate nature of intervention needed. In section five the elementary alternative mathematical structure using Graph Theory is sketched out and applying some algorithms of this mathematics, some implied policy interventions are worked out.

II. Evolutionary Economics

During the last two three decades application of evolutionary thinking to economics of such processes has increased significantly (for an excellent literature survey, please see Van den Bergh *et al.*, 2007 and Nelson and

Winter, 1982,1997). It has resulted from increasing dissatisfaction with the method of understanding technological changes in terms of mainstream economics growth models. Previously taken as exogenous manna from heaven and then being taken as endogenous as they are in neoclassical economics, technological developments along with the changes in institutional environments (that involve policy changes also), are now being viewed as the outcome of continuous interaction between generation (innovation) and selection of diversity in technologies and organizational structures. The traditional behavioural assumptions of individual and group rationality are replaced in evolutionary approach by bounded rationality, which can take the form of habits, routines, myopia and imitation (Robson, 2001). Evolutionary economics, thus gives a realistic alternative to conventional theories of economics and technological change, which includes changes in the technology of management of the economy, that is, economic institutions and policies.

Basic Concepts in Evolutionary Economics

For, charting out an economy in terms of concepts of evolutionary economics one has to look for the elements and process mentioned below which play central and complementary roles within evolutionary economics (as summarized by van den Bergh *et al.*, 2007).

- *Diversity (Variations)*: members of the sets of strategies, products, technologies and organizational structures;
- *Selection*: Processes that reduce existing diversity;
- *Innovation*: Processes that create more diversity;
- *Transferability (transmission)*: replication by reproduction or imitation. This results in possible continuity, durability and cumulative processes;
- *Bounded Rationality*: Individuals and organizations (groups) largely behave according to fixed patterns that result from adjustments in the past, to a certain environment often satisficing that is 'putting out fire' way. Characteristic features of this are selecting survival opportunities under routine behaviour, imitation of others and myopia (short time horizon).

Essentially evolutions comprise of two opposing causal processes. One induces creation of diversity through different mechanisms under discovery and innovations. The new environment visualized by the NAP creates the scope for new agricultural products, machines and inputs, strategic rules for the related agents, are some such members of new diversity.

The other is selection that reduces diversity that combines a number of processes affecting the fitness of individual elements through their survival, reproduction and diffusion. It includes market competition, interaction between employees and employers, relationship with other pressure groups, mergers and acquisitions, financial requirements by providers of capital, legislation and public regulation and public opinion. The interaction of these opposing forces of innovation and selection leads to a process of continuous change, possibly not leading necessarily to an equilibrium. The activities implied by the NAP also rules out the functioning without market penetration and absence of contract farming.

Any new set of policies may be viewed as an organizational innovation that modifies the institutional environment, nature of diversity in possibilities. That is, the choice set, and the ways in which 'selection's would be made. Hence, it can be taken as the institutional counterpart of innovation in a very broad sense of the term.

In this context policies are parts of structural change processes taking place through some mechanism embedded in the selection process and affect the human institutions, technologies and organizations.

III. Evolutionary Economics in Indian Agriculture

There is hardly a single peasant clearly pursuing a smoothly defined neoclassical objective function in Indian agriculture. More difficult it is to find steady-state farmers repeating the same decision because they definitively know this as the optimal under the given circumstances. It is also generally impossible to find small farmers in developing country agriculture not looking for dependable ways out from the current pressures on his/her survival. On the contrary, it is more likely to come across people struggling hard with the inherent uncertainties, and with coevolving pressures, and hence, are 'satisficing' under bounded rationality (Nelson and Winter, 1982). There is some kind of path dependence (lock-ins) which cannot be avoided easily. Besides, nothing stays fixed/ steady as in neoclassical static/dynamic equilibrium.

Indian Agriculture along History Dependence

In reality, the entire agrarian economy is always in a continuous flux from one techno-economic state to another, chunked out through survival strategies of constituent people in the current state. The change works out through and out of decisions of individuals inclined to change for surviving the onslaught of continually emerging adversities including those following from differently oriented policies, as our small cultivators often found to be following. This is better depicted as some evolutionary process.

On the other hand, rather than being a result of some policy intervention from above, economic development has always remained a coevolving process of exclusion and inclusion brought about by evolutionary processes emanating from survival behaviour of the economic agents and institutional decision makers, small farmers being a component of the set. Gradual fragmentation of landholdings over generations of small farmers, their effective bondedness to money lending landlords and hence more and more intensive cultivation of smaller holdings on the one hand and reverse tenancy type reallocation of land in favour of the rich and big farmers is one exclusion aspect of such an evolutionary process. They are further excluded due to degradation of environment a result of technological changes over to large scale processes.

Conventional methods of environmental economics are not adequately equipped to capture such exclusions, a point we discuss in greater details, later on. To delineate these transitions we have to read such trajectories of the economies that involve continual changes of the input combinations that will imply that getting bigger is more profitable. It is not a case of optimization but to remain at least as good as the current state of affairs

One should not assume that this arises out of some choice deliberately made under smooth objective functions and optimization along that line. It is too abstract to provide a policy base so far as the small farmers under distress are concerned. It is necessary that the question is properly asked: why is small farmers' miseries emerging from the process of economic development under the new economic policy; and where exactly the policies are going wrong. We ask these two questions and look for an appropriate analytical framework in the discrete mathematical formulation of such an economy under evolutionary behaviour of the economy and the constituent agents.

Let us check whether we find this evolutionary set of elements in Indian agriculture. While looking for population with diversity of agents, strategies and technological opportunities, we find the following; the coexisting big, medium, small and marginal farmers; sellers of technology, inputs, such as, seeds, fertilizers; evolution of the nature of crop selection (local agents with traditional crops handed over to next generation, by agents family-elders, village power, religion and local traditional wisdom); at the organization aspect, the rising population on limited size of land, gives rise to fragmentation of land holdings over time, threatening family sustainability and contributing to crop and market failures.

The processes ensuing call for external support of savings needing change of mechanisms bringing about the advent of external interest groups, relevance of foreign economic and political power and foreign market interests which generate new diversities and ecological imbalances underneath. This is kind of history of unfolding new diversities. Without this the poor farmers would be working with primitive conventional methods and inputs initially. Inducements to organisational and technological changes were weak while overdue and hence the drive came from the Government as a group decision maker in the system. Dominance and growth of the clutches of the informal money lenders previously turned out to be a strong retrograde factor, adding to the already overwhelming pressure against survival.

Looking for the 'selection', one has to examine the changing organizational features of Indian agriculture over a fairly long enough time since independence. Behaviour of an extremely poverty stricken mass of people along with a few rich farmers, engaged in exploitative institutional relationships with the landlords and informal money lenders were found rampant. Gradually, the working of the system under *panchayati raj* and the organizational changes after the 73rd and 74th Amendments of the Constitution of India reflect changing selection procedures operative in the regimes.

Transferability is found in pursuing certain routine patterns as emanates from the system of governance and the inertial clutches of traditional attitudes, lack of adequate education and research, induce the bounded rationality to be the main behavioural approach to individual and group selections.

Path-Dependent Policy Evolution and India's Small Farmers

Apparently in the policy space, frantic search for organizations free from such oppressive inertial forces were on in the Government's interaction with the pressures while in the affected humans' drives were found to have been expressed in periodical revolts or rebellion of the poor against the power. Only minor changes took place through inert reluctant policy interventions, such as, reducing the land ceiling rather ineffectively, repeated but futile mentioning of land reforms, and/or recognizing the need for decentralization and stray drives for formation of farmers' cooperatives. The search grew appreciably stronger over time. The need for changing the relations and quantities became imperative. Starting with a complacent bur innovative attitude with agriculture being at the centre in the First Five Year Plan, it passed on to industrialization drives along with call for major agricultural changes subsequently. Resultantly changes in the knowledge space and emanating diverse possibilities took place. The need for raising the agricultural productivity was emphasized ever since, as the Green Revolution exemplifies in how it spread starting with the Intensive Agriculture District Programme (IADP) programmes concentrating on selected villages but spread over the North West and allover the country subsequently. The components of the new technology, such as, chemical fertilizers, pesticides, High Yielding Variety (HYV) seeds artificially developed from the laboratory, and associated irrigation arrangements, required that the inherent economies of scale were availed. As a result the new possibilities were virtually opened to big farmers putting the small and marginal farmers into stronger adversities. Under the ongoing threats of food security another Green Revolution of twenty first century is in the offing.

This was accompanied by a second layer of evolutionary selection process. As noted by many economists and social activists, Green Revolution imposed very adverse environmental effects on soil quality, water stock and the domestic life support system of the pre-existing biodiversity (Reddy, 2011). The entire lost horizon was supportive to small farmers. Hence as the new set of 'diversity' unfolded, it aggravated the stresses of the small farmers further and a negative 'selection' ensued for them in the sense of reverse tenancy or rampant suicides. The process has been fast dislocating small and marginal farmers from regular agriculture.

Following Ramasubba Reddy we can see the plights of the small and marginal farmers being 84 per cent of a total in 2001 census. Their earnings are abysmally low and inadequate for a proper subsistence. The situation has been worsening, because of risks not only from nature but from the caprices of increasing dependence on markets. Small farmers do not have the strength to absorb the shocks therein. Rising costs of cultivation, low remunerations, high risks with frequent crop failures, declining agricultural growth, and mounting debts have all led the farmers to serious distress.

The entire crop of the small holders comes to the market at one time. The small cultivator, who is often heavily indebted, has poor bargaining strength and gets a raw deal from the more resourceful traders. Nearly 40 per cent of the farmer households disliked their occupation because of non-viable state of agriculture for the small and marginal farmers. Inadequate insurance and minimum price support from the government further deepens their sufferings.

Increased indebtedness is noted as a major reason for the spurt in farmer suicides during recent times while the economic status of the suicide victim was very poor, being small and marginal farmers. Among the reasons for suicides, indebtedness featured as the prime reason. After the Green Revolution agricultural activities have become cash based individual enterprises requiring high investment in modern inputs and wage labour .This is evident from the list of states with high incidence of farmer suicides, which are not necessarily backward or predominantly agrarian or with low income.

Taking the words of Reddy, "Increased liberalization and globalization have in fact lead to a shift in cropping pattern from staple crops to cash crops like oilseeds and cotton, requiring high investment in modern inputs and wage labour, and increasing credit needs but when the prices declined farmers had no means to supplement their incomes. When crops failed and/or prices went down they had no means to repay the loans, which drove them to the wall, as the saying goes. Desperate as they were and further burdened by a sense of social shame, ending one's own life might have provided an easy exit. With mounting debt burden along with the rising risks in production and price fluctuations leading to low remuneration, it is no wonder that a lot of distress is generated among the farmers. Incidence of indebtedness among farmer households was the highest in Andhra Pradesh (82%), followed by Tamil Nadu (75%) and Punjab (65%)".

The new threat is the Second Green Revolution which on the other hand has proved worse further. The major thrust of the new revolution is again on the new GM seeds like HYV's of the first. They come with associated non-land input requirements which do not favour small farmers at all. New crops have to be ventured. Whether real net benefit from this agriculture is significantly positive, considering both private and social dimensions is yet to be convincingly proved. A point is clear, however, that the new revolution favours capitalist agriculture which goes against small farmers. All information received so far indicates that it involves higher burden in the cost side, though net benefit may be greater. From two recent works it can be inferred that (a) size of holding is positively related; and (b) the benefit is highest, if cultivation is mono-cropping on the new variety (Ramakumar and Raut, 2011) and (Swaminathan and Rawal, 2011). Unqualified cannot be landed to a claim of higher net benefit if small farmers are taken into account. But cotton cultivation in Maharashtra indicates that there is a strong possibility that the small farmers have been losing their land to the big in exchange of employment as agricultural labourers . The new cultivation is extremely risky for the small farmers and quite costly to ascertain the nature and authenticity of the seeds (Roy, Herring and Gleiser, 2007 and Chandrashekhar and Deb, 2009).

These developments are not compatible with Neoclassical Economics of profit maximization. Starting with assumed unfettered pursuance of optimization leading to equilibrium and steady-states over time is likely to generate increasing well-being tautologically which is never quite found in reality. On the one hand produced commodities do not have smooth production functions nor are there smooth and continuous objective functions surviving the Inada conditions in reality. Secondly, the economics concerned cannot capture the transition to concentration of power. Also the major technological changes are often discontinuous and involve local technological adjustments not amenable to the standard neoclassical tool box of analysis. Rather the existing diversity is easily captured by discrete mathematics and working of a realistic assumption of 'bounded rationality' can capture the evolutionary dynamics that we describe. This chapter formulates such a model in the next section.

One must note that over the post-independence period the changes described above has completely changed the economic, institutional, and societal inter-linkages and networks that constitute the political economy of India.

A Description of an Economy in Evolution using Discrete Mathematics

Attempts to analyze economic phenomena using discrete mathematics and combinatorics are not new. Apart from applications of simple graph theory models to problems of microeconomics there have been applications to input-output models and operations research. Discrete mathematics has been applied to incorporate many other unconventional assumptions in economics also (Roberts, 1976, 2000; Potts, 2000) has depicted a considerably comprehensive correspondence between evolutionary economics and network theory concepts. According to him economic systems are complex hyper structures of nested sets of connections between separate components. This may be components of technology (cars, computers, machines) and/or material and communication flows between individuals or departments in an organization. The notions of business routines, as suggested by Nelson and Winter are also interaction between individuals with specific skills and are also included in the superstructure. The other side of the story is there are elements from a past hyper structure which get excluded or disadvantaged in the present or future hyper structures resulting through the selection process.

In this context all changes in the economic systems, in policies, growth, technology or knowledge are essentially processes of change in this connection. Along this line Potts and also Roberts suggested a new approach to analyses based on discrete mathematics. Potts pointed out the uses of already available multi-agent evolutionary models, known as 'artificial life' or artificial world simulation models. This chapter proposes a graph theoretic model with simple properties robust enough to capture policy changes and their effects in a manner such that appropriate policy change requirements can be adequately anticipated and effected with success rather than failures.

Graph Theoretic Presentation of an Evolutionary Economy

It is well known that as in Sraffa and Leontief, economies can be presented by inter-industry relationships using input-output (I-O) systems. Taking the array of industries and labour with final consumption, in a square matrix one can represent all transactions in terms of input-output flows consistently. What is relatively less known is that any type of returns to scale can be embedded in such systems, the input coefficients, though fixed *ex post*, are functions of prevailing scale economies.

Every I-O structure of an economy can be represented using Graph Theory. Each vertex may represent an industry (economic activity) with edges representing I-O interdependence. There will be an edge from vertex i to vertex j if a_{ij} is nonzero. Digraphs would indicate the direction of dependence and their weights would represent quantities like, value of the flow, cost of the flow or any other quantitative measure associated with such values. Thus by such simple representation the total value of a structure and its distributional implications are discernible from the Graphs.

The upshot of the features of evolutionary economics taken together is that there is always considerable history dependence in policy that itself is a selection from a limited set of diversity. In the technical literature on the bearing of evolutionary economics on economic policy, this is known as path dependence. Essentially this implies that any selection by the economy or its members follow from a multiplicity of open possibilities (may be approached using non-ergodic stochastic dynamics) leading further to a multiplicity of new diversity set (for an excellent comprehensive survey in relation to policy formation, please see David, 2000). Though this a very complicated choice behaviour, but it is characterized by two basic properties: bounded rationality and validity of principle of least action. Policy having chosen its concern must minimize the cost associated with the adverse outcomes.

In the present context, an approach to the selection of policy will be to reduce losing of small farmers' interest more and more. If it is possible to infer the costs of losing small farmers' survival over time for each selected policy mix in this regard, then every economic structure reflected in input-output transactions corresponding to a particular policy can be represented by a valued vertex with weighted edges stemming out indicating the costs of all policy alternatives. To the extent these cost assignments to possibilities are close to reality, bounded rationality is likely to trace a path-dependent chain that will imply minimum total net cost or benefits over time in terms of losses to the small farmers. The technical appendix at the end, proves that the shortest path algorithm makes the choice scenario computable and hence an evolutionarily efficient 'natural' selection is inducible through appropriate evolutionary policy choice by the government or any other relevant agencies.

Translating an Input-Output System into a Weighted Digraph

It is now clear that any I-O structure can be represented by Graph G, (please see the Graf-figure in the appendix) where w_{ij}'s would be replaced by a_{ij}'s, the conventional input coefficient meaning requirement of i-th industry for one unit of j-th output. Starting with a given resource-knowledge-technology base with embedded change forces, such a structure not only have very wide range of 'diversity' underlying, but is capable of capturing a wide possibility of changes. And weights can capture among other things the rate of survival of the entire economy, of its components presentable in terms of partitions defined accordingly or presented by some associated Graph separately. There is no reason why variable returns to scale cannot be tractable in this frame if such a digraph has a dynamic counterpart, as there generally will be a related Graph representing say the investment allocations in terms of the commodities and even growing (with positive or negative growth rate) labour requirements at the vertices, can be represented by another associated weighted digraph. It may also be used to represent land allocations to the nodes along similar lines as labour. Hence if W(t) is taken as the total value of the weights of such a Graph, conceptually W(t) may represent, for example value of minimum-size holding of land, or land/output ratios, minimum profit, incremental reallocation of land to the vertices, population growth in different sectors and so on, with possible W(t + 1) implications.

Now, the existing I-O structure has to be taken as resulted through choice under given constraints with bounded rationality. As it represents reality, the structure has dynamic implications and nodes representing transfer function parallels. It results, in terms of Evolutionary Economics terminology, from a selection process which remains operative to determine the subsequently emanating I-O structures.

Shortest Path Algorithm on the Grand Graph

The specific I-O structure operational at a point of time is the one that has been realized through multidimensional 'selection' process along evolutionary developments over time. This, in turn, is embedded with changed contents, base and a process of change. As a result a new selection process is immediately active as soon as the one realized takes place. Variable weights on industrial processes along time sequence is also a sign of tractability of variable returns to scale, increasing, constant or decreasing. All kinds of changes in the quantitative aspects of an economy can be analyzed using Graph Theory without requiring a necessary continuity assumption on the objective functions or the on technology or anywhere, so to say.

1. 'Satisficing' with bounded rationality rules out all edges with increased cost measured weights per unit of output. Hence selection either reduces cost or keeps it fixed at the current level. The first layer of selection will be on a chain.
2. The rule operative at the layer ends the search and generates the set of all such chains which will be time subscripted (see figure 3.2 in Technical Annexure 3.1) to indicate the I-O of the

operational phase. And the implied other costs such as biodiversity or environment works through a sub-phase since they have to be taken care of during the phase only.

3. Each time subscripted edge has at least two values as weights, one for the first layer and second for the implied sub-phase layer.
4. Selection through shortest path chain will be complete after both the layers have worked out.

The above mentioned process of selection implies the following propositions (proofs are simple and skipped).

Proposition 1: First layer selection along shortest path chain implies availing economies of scale provided they are available.

Proposition 2: After the first layer selection all other first layer possibilities will be rejected, as they are irrelevant.

Proposition 3: The remaining weighted digraphs are either possibility in the second layer (there will be a second layer as there will be cost implying sub-phases to be taken care of immediately) till that period and the first layer possibilities from that phase onward. Selection process at both levels remains operative till completion of the shortest path chain.

Proposition 4: All second layer selections will avail economies of scale provided they are available.

Proposition 5: Satisficing with economies of scale will benefit big agents and rejects smallest agents.

As we have stated earlier, the algorithm is applicable for all possible partitions of the graph, it is possible to partition the graph along the nature of output, agriculture and non-agriculture. Then the satisficing at the first layer will either reduce cost, or keep it fixed per unit of output. Hence those farms will survive for whom this is true. Cost will reduce possibly for those who increase in size, as needed by the first green revolution implications. For those who survive with cost at the first layer fixed are smaller ones and gets disadvantaged by the relative scale advantages for those with fallen costs. Those who cannot keep costs fixed either will not survive.

Many smaller farmers are likely to face extinction at the second layer because of the adverse effects of environmental degradation of the life support system.

IV. How does Policy Environment Benefit from this Formulation?

As we have stated earlier, the algorithm is applicable for all possible partitions of the graph, it is possible to partition the graph along the nature of output, agriculture and non-agriculture. Then the satisficing at the first layer will either reduce cost, or keep it fixed per unit of output. Hence those farms will survive for whom this is true. Cost will reduce possibly for those who increase in size, as needed by the first green revolution implications. For those who survive with cost at the first layer fixed are smaller ones and gets disadvantaged by the relative scale advantages for those with fallen costs. Those who cannot keep costs fixed either will not survive.

Many smaller farmers are likely to face extinction at the second layer because of the adverse effects of environmental degradation of the life support system.

We have noted that starting from initial node, possibilities are discrete and finite. We also noted that the graphs, weights, value of weighted digraphs and shortest path chains are all computable and amenable to simulations. The policy authority then easily can simulate anticipated shocks of all kinds in this framework and resulting policy effects can be exactly anticipated. The policy can work out its own theoretical structure that may assign some predictive philosophical strength at the base.

Vi. Conclusion and Policy Implications

The description and the model above reflect a process that reflects agents to be continuously under pressure for survival at all points of its cohesion. As a result, a 'satisficing' behaviour at every stage results. As a result, survival depends on growth of implied capital or land holding, commensurate with the availability of scale economy cost advantages. For survival of the farmers it would be necessary to increase in size which is possible only if land (inelastic in total supply) is redistributed in favour of bigger size. Small farmers are at the losing trajectory.

For policy implication it is suggested that the 'selection' is not a natural one for our agriculture (Shiva, 1999). The technology for survival emanated from alien interest and foreign ecology. The destruction of agricultural resource base with green revolution is a proof of the anomaly leading to serious environmental consequences. Policy has to look for action research and intervention that brings back our own seeds and technology that consistently supports both small and big farmers on an equal plane. For this purpose doing away with the dependence on smooth optimization-based neoclassical understanding in favour of realistic discrete mathematical analyses is likely to generate stronger framework for policy formulation.

REFERENCES

David, Paul A. (2000). "Path dependence, its critics and the quest for 'historical economics'" in P. Garrouste and S. Ioannides, eds. *Evolution and Path Dependence in Economic Ideas: Past and Present*, Edward. Elgar Publishing, Cheltenham, England.

Jacquet, P.; R. K. Pachauri, and L. Tubiana. (eds.) (2011). Development, the Environment and Food: Towards Agricultural Change? Afel, IDDRI, and TERI.

Kumar (2006). "Contract Farming through Agribusiness Firms and State Corporations: A Case Study of Punjab" *Economic and Political Weekly* December 30.

Ministry of Agriculture, Government of India, NAP (2011).

Narayanmurthy, A. (2006). "State of Indian Farmers" *Economic and Political Weekly,* February 11, 2006.

Nelson R. R. and S. Winter. (1982). *An Evolutionary Theory of Economic Change,* Harvard University Press, Cambridge, MA, USA.

Potts, J. (2007). *The New Evolutionary Microeconomics: Complexity, Competence and Adaptive Behaviour*, Edward Elgar, Cheltenham, U.K.

Ramakumar, R. and Karankumar Raut. (2011). "Agricultural Employment in a Vidarbha Village: Results from a Resurvey" *Review of Agrarian Studies*, Vol.1, No.1 pp. 63-100.

Rao, N. Chandrashekhar and S. Mahendra Dev. (2009). "Biotechnology and Pro-Poor Agricultural Development". *Economic and Political Weekly.* 46, (52). Dec. 26, pp. 56-64.

Reddy, B.S. (2011). "Dynamics of Soil Fertility Management Practices in Semi-arid Regions: A Case Study of AP". *Economic and Political Weekly,* January 15.

Roberts, F. (2001). "Some Applications of Graph Theory" Mimeo Department of Mathematics and DIMACS, Rutgers University.

Roberts, F. (1976). *Discrete Mathematical Models with Applications to Social, Biological and Environmental Problems,* Prentice Hall, Englewood Cliffs, N.J.

Robson, A.J. (2001). "The Biological Basis of Economic Behaviour", *Jouurnal of Economic Literature,* 39, pp. 11-33.

Roy, Herring and Gleiser (2007). "Naturalising Transgenics: Official Seeds, Loose Seeds and Risk in the Decision Matrix of Gujarati Cotton Farmers" *Journal of Development Studies*, Vol. 43(1), pp. 158-176.

Shivaa, V. and Ashok Emani, Afrar H. Jafri (1999). "Globalization and Threat to Seed Security: Case of Transgenic Cotton Trials in India" *Economic and Political Weekly,* March 6-13.

Sun, L. and M. Wang, "An Algorithm for a decomposition of weighted digraphs – with applications to life-cycle analysis in ecology" Mimeo (Website sited in the paper.)

Swaminathan, Madhura and Vikash Rawal (2011). "Are there Benefits from the Cultivation of Bt Cotton? A comment on data from a Vidarbha Village" *Review of Agrarian Studies*, Vol.1, No.1 pp. 101-124.

Van den Bergh, J.C.J.M.; A. Faber; A.M. Idenburg and F.H. Oosternhuif (2007). *Evolutionary Economics and Environmental Policy*, Edward Elgar, Cheltenham, U.K.

Vyas, V.S. (2004). "Agrarian Distress: Strategies to Protect Vulnerable Sections" *Economic and Political Weekly,* December 25.

Wilson, K. (2002). "Small Cultivators in Bihar and New Technology Choice or Compulsion" March 30.

Technical Annexure 3.1

Graph theory Terminology (from Sun and Wang, *Journal of Mathematical Biology* http://spingerlink.com)

A directed graph G is described as $G = (V, E)$, where $V = (\nu_1, \nu_2, \ldots\ldots, \nu_n)$ is the set of *vertices* or *nodes* of the graph G. $E = \{ e_{ij}, = (\nu_i, \nu_j), 1 = i, j = n\}$ is the set of directed edges or *arcs* of G that goes from vertex ν_i, to ν_j, the ordered pairs of vertices of G. $|E(G)|$ is the total number of edges of the graph G. A weighted digraph (G, w) is a digraph G with a numerical weight assigned to each directed edge and can be represented by a matrix as given below:

From vertices

$$\text{to} \quad \begin{matrix} & \nu_1, & \nu_2, & \ldots, & \nu_n \\ \begin{matrix}\nu_1\\ \nu_2\\ \nu_n\end{matrix} & \multicolumn{4}{l}{\begin{bmatrix} w_{11}, & w_{12}, & \ldots, & w_{1n} \\ w_{21}, & w_{22}, & \ldots, & w_{2n} \\ w_{n1}, & w_{n2}, & \ldots, & w_{nn} \end{bmatrix}} \end{matrix}$$

where w_{ij} is the weight of the directed edge from vertex ν_i to ν_j.

A chain or a simple chain is a trajectory of a sequence of edges in G that connects nodes $\nu_{i1}, \nu_{i2}, \ldots, \nu_{ik}$, consecutively, where $\nu_{i1}, \nu_{i2}, \ldots, \nu_{ik}$ are distinct. A graph is called connected if any two of its vertices are linked by connected edges, that is, by a *path* in G.

In this chapter only connected graphs which are robust to certain theorems that allow graph correspondences and spanning of associated graphs.

From such (V, E) a weighed digraph $G(w)$ is always derivable by assigning weights and then the total value of weights, value of weights of all feasible partitions of (V, E) and so of $G(w)$ are derivable. Then from each such graph, it is possible to figure out further sets of implied weighted digraphs with weights having been changed accordingly. For example, from the initial $G(w)$, the value of direct and indirect labour requirement at each node may be a parallel array of weights for the same graph.

Similarly if the scale of operation at each node is indicated by a certain parametric value at the spanning base scale values can characterize another weight matrix for the same initial graph. So can be the values of implied biodiversity or environmental cost at each node and for each such transformation of the initial graph,. The total value of the weights for the entire graph and all its partitions can be derived. Hence from a basic information-knowledge-material space, one I-O weighted digraph and all its implied weighted digraphs are computable and amenable to useful policy simulations.

For addressing dynamic aspects, that is time dependent choice, more consolidated envisioning is possible. Take any graph (G, w), as the vertex itself including its knowledge implications, $A_{gi}(w)$, i-value indicating the stage identity of the Graph on the trajectory implying the time-chain on the graph. w is the total value of that graph. Thus from a graph generated initially, move to a set of graphs over time to some intermediate or final stage, denoted by a grand graph $\Omega\ (t)$ is derivable applying certain theorems of Graph theory. Such an $\Omega\ (t)$ is capable of subsuming all possible histories of evolution of selection till time t. Choice over the $\Omega\ (t)$ is also guided by Shortest path algorithm under bounded rationality with possible path-dependences (lock-ins).

Take the following I-O system of three commodities, for example:

	x_1,	x_2,	x_3	C
x_1	x_{11},	x_{12},	x_{13}	C_1
x_2	x_{21},	x_{22}	x_{23}	C_2
x_3	x_{31},	x_{32}	x_{33}	C_3

which can easily be represented by the standard Leontief system notations?

	x_1,	x_2,	x_3	C
x_1	$a_{11}x_1$,	$a_{12}x_1$,	$a_{13}x_1$	$c_{11}x_1$
x_2	$a_{21}x_2$,	$a_{22}x_2$	$a_{23}x_2$	$c_{22}x_2$
x_3	$a_{31}x_3$,	$a_{32}x_3$	$a_{33}x_3$	$c_{33}x_3$

We can easily translate it into a graph, taking column heads as the vertices and whenever there is nonzero a_{ij}'s we say there is an edge from i-th vertex to the j-th. The direction of the flow determines the direction of the edge, the digraph of that part. a_{ij} can serve the purpose of a weight or any other quantity implied by them or may be even externally assumed associated quantities may be assigned weights to the edges. The I-Os may be defined over intervals representing the range of commodities out of the larger set of producible and thus may be used in intermediate good relationships also.

Let A_1 be a I-O matrix-spanned graph. Weighted digraph $A_1\ (w)$ is found by taking a_{ij}'s as the first level of weights. Cost-weighted digraph $A_1\ (w_c)$ can then be found by taking the cost correspondent of the a_{ij}'s.

Thus the basic I-O system can be represented using any one of the related weighted digraphs. One has to note that any A_i is generated from a knowledge base K_{i-1} and leads to a new one K_i. The rule of 'selection' in this and subsequently resulting I-O systems are guided by the universal least action principle which in terms of graph theory takes the form of shortest path algorithm.

Initial digraph A_1 through its new knowledge base often generate diversity sets P_{Aij}'s from which the selections are made of the subsequent A_{2k} possibilities. (Figure 3.1 below):

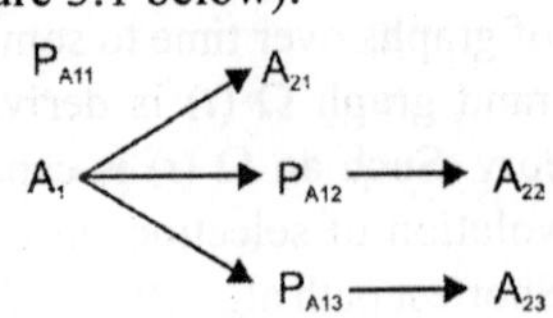

Fig. 3.1: First layer of the two period selection process: Possible Graphs resulting from A_1

P_{A1i}'s represent the *i*-th possibility set of activities and networks of digraphs, leading to the possible selection of graph in the second period A_{2i} through shortest path algorithm. One and only one A_{2i} will result. From each such A_{2i} through P_{A2j}'s a set of A_{3j}'s will result and each edge from A_{2i} to next A_{3j} will carry the respective weight of a certain cost aspect of the choice (Figure 3.2 below).

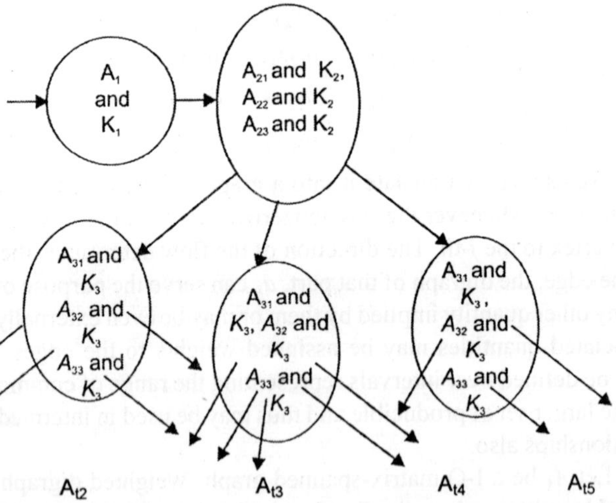

Fig. 3.2: Grand Graph G with $|E(G)|$: Weighted digraphs in sequence on knowledge base K_j

However, since no continuous objective function is necessarily present, optimization is replaced by selections through satisficing, under 'bounded rationality', that is, ensuring survival of the best possible kind rather than maximizing size or any such expansionary attribute.

Start with a connected digrapgh G with $|E(G)|$ directed weighted edges.

Consider a possible simple chain that moves along the nodes of 'incidences'. This will be a *cycle* if the chain ends at the initial node.

Extract a simple chain without contradictory directions from G, by searching among all simple chains.

A simple chain starts and ends at different nodes A_1 to A_{i1} with the condition that $w_{i,1}>0$ (weight for the edge between A_1 and A_{i1}) for some $i \neq 1$ (there is at least one edge from A_1). Search out through the rest nodes A_2 to A_n choose the first i_1 such that $w_{i1,\ 1}> 0$ (edge A_1 to A_{i1} exists). The path selection may be stated again as follows:

(i) 'Satisficing' with bounded rationality rules out all edges with increased cost measured weights per unit of output. Hence selection either reduces cost or keeps it fixed at the current level. The first layer of selection will be on a chain.

(ii) The rule operative at the layer ends the search and generates the set of all such chains which will be time subscripted (see figure 3.2) to indicate the I-O of the operational phase. And the implied other costs such as biodiversity or environment works through a sub-phase since they have to be taken care of during the phase only.

(iii) Each time subscripted edge has at least two values as weights, one for the first layer and second for the implied sub-phase layer.

(iv) Selection through shortest path chain will be complete after both the layers have worked out.

4

Adoption of Innovative Agricultural Practices across Supply Chain among Small-Scale Vegetable Growers

Jabir Ali and *Tribhuvan Nath*

Introduction

The food and agricultural sectors in developing countries have significantly been getting transformed, where the relative importance of grains and staple foods are declining while that of high value agriculture such as fruits, vegetables, milk, meat and eggs are significantly increasing (Gulati *et al.*, 2007). As most of the high value agricultural products are comparatively perishable in nature, it requires greater coordination in the way the food is produced, processed, marketed and consumed (Deshingkar, *et al.*, 2003; Busch and Bain, 2004; Henson and Reardon, 2005; Swinnen and Maertens, 2007). Due to structural changes in global food consumption and demand, the productions of these high value commodities are also responding in the similar manner. India is the second largest producer of vegetables in the world with over 122.3 million tonnes after China and accounts for about 15 per cent of the world's production of vegetables. Around 7.7 million hectares of area is under vegetable cultivation, which is about 5.4 per cent of the total area under cultivation in the country. The vegetable production in the country has significantly increased in the recent decade with the advancement and the adoption of new techniques conducive to different soil and climate types.

Emerging consumer markets for vegetable and vegetable products need new direction on cultivation practices, high yielding and modern variety seeds, fertilizers, pesticides and insecticides, irrigation technologies, post-harvest management and logistics and distribution for the production of vegetables. In developing countries including India, most of the vegetables supplied to the market are produced by small scale farmers, using traditional method of farming, immediate intervention for improving the cultivation practices become more important. Although the future of vegetable production in the country seems very bright, however for realizing the full potential of the vegetable sector, efforts need to be made in delivering innovative practices/knowledge among the vegetable growers for sustainable development. Empirical evidence suggests that farmers with the resources and skills to produce high value commodities are increasingly getting better opportunities for increasing their income and employment (Birthal *et al.*, 2005).

This study analyzes the adoption behaviour of small-scale vegetable growers on innovative farm practices by across supply chain-cultivation, post-harvest practices and marketing of vegetables-based on a primary survey of 556 vegetable growers in six districts of Uttar Pradesh (India). The study has far reaching impacts on designing and implementing policies related to vegetable value chain. The results will also enhance understanding of the adoption of innovative agricultural practices for targeting the agrarian communities in a better way.

Data and Methods

Data Collection

This study is based on the primary survey of 556 vegetable farmers belonging to six districts of Uttar Pradesh in India, viz., Lucknow, Allahabad, Gorakhpur, Moradabad, Jhansi and Agra. A survey schedule was canvassed personally and farmers' responses to various questions related to adoption of agricultural practices were obtained and recorded. A survey was conducted using pre-tested structured schedule and responses on various agricultural practices were asked through personal interviews. The survey schedule was developed in Hindi language to ensure greater participation of the respondent in the survey process on one hand and to increase the confidence of the respondent in providing accurate and unbiased information on the other. The schedule was broadly divided into three parts comprising data on socio-demographic profiles of the respondents; farming practices; and sources of information used for vegetable cultivation decision making.

Data Analysis

Simple statistical analysis has been undertaken such as cross-tabulation, percentage and chi-square tests, using Software Package for Social Science (SPSS) version 20.0. Adoption of innovative agricultural practices by the vegetable growers has been divided into three supply chain stages – cultivation, post-harvest management and sales and marketing– comprising specific indicators (Figure 4.1). To assess the difference in adoption behaviour between small landholders (<2 hectares) and large landholders (> 2 hectares), chi-square statistics has been used for each indicators across the vegetable supply chain.

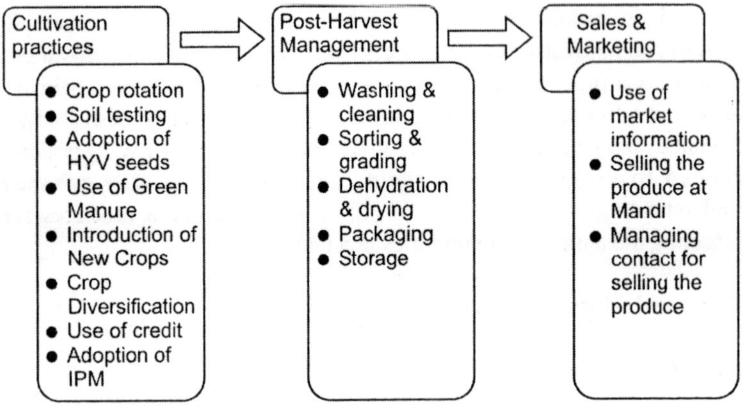

Fig. 4.1: Adoption of agricultural practices across supply chain

Further, based on literature review, an empirical model has been developed to identify the factors affecting farmers' adoption of innovative agricultural practices among small vegetable growers. The adoption of innovative agricultural practices is considered as a dependent variable whereas socio-economic factors such as age, education, income level, social category, farm training and the farm related factors such as landholdings, area under irrigation, number of crops grown, and hiring of labour are considered as independent variables.

The Poisson Count Regression Model (PCRM) is used to analyze the factors influencing the adoption of agricultural practices by small vegetable growers. A generalized linear model is Poisson if it follows Poisson distribution and the link function is log. Maximum likelihood approach was adopted in estimating the efficiency of the regression model. A Poisson model of a count variable is estimating the log of the expected count as:

$$\log \lambda_1 = \alpha + \beta_i X_i + \varepsilon_i \qquad \text{...(1)}$$

Where, log λ_1 is expected value of the dependent variable for the *i*-th observation, α is constant, β_i is parameter estimates of demographic and farm level factors that influence adoption, denoted by the vector *Xi* and ε_i is error term. The model assumes that the dependent variable results from a counting of events using positive integer numbers. In this model the regression coefficient β_i represents the expected change in the log of the mean per unit change in the predictor X_i. In the model, an exponentiated regression coefficient $\exp^{\beta}$ represents a multiplicative effect of the *i*-th predictor on the mean, where increasing X*i* by one unit multiplies the mean by a factor $\exp^{\beta}$.

Table 4.1: Variable descriptions and summary statistics

Variables	*Description*	*Mean*	*Std. Deviation*	*Expected Sign*
Dependent variables				
TOTCULT	Adoption of cultivation practices	2.88	1.32	
TOTPOSTHR	Adoption of post-harvest practices	1.65	1.22	
TOTSLMKT	Adoption of sales & marketing practices	0.79	0.71	
GRTOT	Overall adoption of innovative practices	5.33	2.10	
Independent variables				
AGE	Age of Respondents	40.33	15.67	–
EDU	Education (1 if educated, 0 otherwise)	0.28	0.45	+
SOC	Social category (1 if OBC, 0 otherwise)	0.65	0.48	–
INC	Monthly household Income (1 if >5000 pm, 0 otherwise)	0.37	0.48	+
WRKMB	Total Working Members	3.60	2.33	+
LANDSZ	Total land ownership	1.71	2.33	–
LEASED	leased-in land (1 if leased-in, 0 otherwise)	0.32	0.47	–
IRRG	Irrigated area in percentage (1 if >50%, 0 otherwise)	0.81	0.39	+
ICT	No. of ICT uses	0.42	0.92	+
CROPS	Total no. of crops	5.28	2.46	+
DISTANCE	Market distance	0.70	0.46	–

Based on the variable used in the present study, the empirical model was specified and estimated to predict the likelihood or probability of the factors influencing adoption of innovative practices by the vegetable farmers, as follows:

$$\text{Log } \lambda_i + \alpha + \beta_1 AGE + \beta_2\, EDU + \beta_3\, SOC + \beta_4\, INC + \beta_5\, WRKMB + \beta_6\, LANDSZ + \beta_7\, LEASED + \beta_8\, IRRG + \beta_{9ICT} + \beta_{10}\, CROPS + \beta_{11}\, \text{DISTANCE} + \varepsilon_i \qquad \ldots(2)$$

The description of variables used in the model is given in Table 4.1. The expected directional effects of each independent variable are also indicated. The results of a Poisson regression model can also be described in terms of a rate ratio or 'relative rate'. This model can be used to predict the expected level of adoption by exponentiating the coefficient value of each independent variable, while assuming the effect of other variable as constant. Hence, the percentage change on λ_i due to each independent variable X_i can be calculated as follows:

$$\Delta\lambda_i = 100 \times (\exp^{\beta} - 1) \qquad \ldots(3)$$

Results and Discussion

Profile Analysis of Adopters and Non-Adopters

Table 4.2 provides summary profile of the innovative agricultural practices adopters and non-adopters sample farmers with respect to age, education, social category, household income and landholding size. Out of 556 vegetable growers, 400 respondents were reported the adoption of innovative farm practices across vegetable supply chain, i.e., 72 per cent of total vegetable growers. The majority of the respondents were in the age between 20 years to 60 years, indicating a mature group involved in vegetable cultivation. However, the average age of both adopters and non-adopters estimated to be 40 years, implying that both the groups have same level of farming experiences ($\chi^2 = 3.404$, $p = 0.333$). The results of chi-square tests revealed no significant difference in education levels between the adopters and non-adopters ($\chi^2 = 10.327$, $p = 0.243$). Generally, educated farmers are more likely to adopt innovative agricultural practices across the chain. As for as social groups are concerned, majority of both the adopters belong to socially backward groups ($\chi^2 = 1.469$, $p = 0.689$).

Table 4.2: Sample demographic characteristics

Socio-demographic variables	*Overall (N=556)*		*Non-adopter (N=155, 28%)*		*Adopter (N=400, 72%)*		*Chi-square*	*Sig.*
	N	*%*	*N*	*%*	*N*	*%*		
Age								
<20 yrs.	55	9.9	12	7.7	43	10.8	3.404	0.333
20-40 yrs.	260	47.0	82	52.9	178	44.7		
40-60 yrs	179	32.4	45	29.0	134	33.7		
>60 yrs.	59	10.7	16	10.3	43	10.8		
Education								
Illiterate/Literate	155	28.2	38	24.7	117	29.5	10.327	0.243
JHS and below	156	28.4	47	30.5	109	27.5		
Sec./H.Sec.	165	30.0	52	33.8	113	28.5		
Graduate/PG	74	13.5	17	11.0	57	14.4		
Social Category								
General	119	23.2	38	26.0	81	22.0	1.469	0.689
OBC	361	70.2	98	67.1	263	71.5		
SC/ST	34	6.6	10	6.8	24	6.5		
Occupation								
Farming	507	92.7	142	92.8	365	92.6	2.859	0.826
Service	14	2.6	4	2.6	10	2.5		
Business	1	0.2	0	0.0	1	0.3		
Others	25	4.6	7	4.6	18	4.6		
Household Income (monthly)								
<Rs. 2000	139	25.4	69	45.1	70	17.7	50.453***	0.000
Rs. 2000-5000	201	36.7	52	34.0	149	37.7		
Rs. 5001-10000	108	19.7	17	11.1	91	23.0		
Rs. 10001-15000	39	7.1	7	4.6	32	8.1		
Rs. 15001-25000	33	6.0	6	3.9	27	6.8		
>Rs. 25000	28	5.1	2	1.3	26	6.6		
Landholdings								
Small holders (<2 ha)	427	76.9	123	79.4	304	76.0	0.709	0.234
Medium & Large (>2 ha)	128	23.1	32	20.6	96	24.0		

***significant at the 0.01 level, **significant at the 0.05 level, *significant at the 0.10 level.

N = Number of observations.

In regard to income, both the adopters and non-adopters of the innovative agricultural practices belong to lower income groups (χ^2=50.453, p = 0.000). Regarding land ownership, average landholdings of adopters was estimated to be 1.67 hectares relatively lower than non-adopters (1.781 hectares). Further, landholdings distribution analysis confirms that the non-adopters comprised of higher proportion of Small holders (<2 hectares) with relatively non-adopters. The results of chi-square tests between the adopters and non-adopters with landholding size indicated no significant difference (χ^2 = 0.709, p = 0.234). The results suggest that the two categories of vegetable growers (adopters and non-adopters of innovative farm practices across supply chain) exhibited statistically significant difference with respect to only income level (Table 4.2).

Adoption of Agricultural Practices across Supply Chain Cultivation Practices

Vegetable growers are required to adopt innovative cultivation practices to maximize the total output in sustainable manner. The objective of increasing production is desirable not only from the individual farmer's perspective but also from the broader societal perspective. The adoption of innovative production technologies comprise a bundle of activities

Table 4.3: Chi-square statistics on adoption of cultivation practices by farm size

Cultivation practices	*Overall*		*Small holders*		*Medium & Large*		*Chi-square*	*Sig.*
	N	*%*	*N*	*%*	*N*	*%*		
Crop rotation								
Adoption	347	62.5	279	65.3	68	53.1	6.270***	0.009
Non-adoption	208	37.5	148	34.7	60	46.9		
Soil testing								
Adoption	65	12.3	50	12.2	15	12.8	0.036	0.479
Non-adoption	463	87.7	361	87.8	102	87.2		
HYV Seeds								
Adoption	422	76.0	317	74.2	105	82.0	3.282**	0.043
Non-adoption	133	24.0	110	25.8	23	18.0		
Green manure								
Adoption	47	8.5	32	7.5	15	11.7	2.267*	0.095
Non-adoption	508	91.5	395	92.5	113	88.3		
Introduction of new crops								
Adoption	181	32.6	138	32.3	43	33.6	0.073	0.433
Non-adoption	374	67.4	289	67.7	85	66.4		
Crop diversification								
Adoption	332	59.8	256	60.0	76	59.4	0.014	0.493
Non-adoption	223	40.2	171	40.0	52	40.6		
Use of farm credit								
Adoption	188	33.9	133	31.1	55	43.0	6.144***	0.009
Non-adoption	367	66.1	294	68.9	73	57.0		
Use of IPM								
Adoption	19	3.4	16	3.7	3	2.3	0.587	0.327
Non-adoption	536	96.6	411	96.3	125	97.7		

***Significant at the 0.01 level, **Significant at the 0.05 level, *Significant at the 0.10 level.

such as crop rotation, soil testing, use of high yielding seeds, use of green manure, introduction of new crops, crop diversification, use of farm credit and use of integrated pest management (IPM). Though, successive agriculture policies in India have been emphasizing adoption of improved cultivation technologies, however, the majority of cultivation practices till date remain subsistence oriented because of many constraints.

Table 4.3 presents the chi-square statistics on adoption of cultivation practices by farm size. The highest rate of adoption obtained for HYV seeds followed by crop rotation, crop diversification and least for IPM. Crop rotation practices for better productivity and sustainability is adopted by 63 per cent of the respondents. The results of chi-square tests revealed significant difference in adoption levels between the small holders and medium and large land holders ($\chi^2 = 6.270, p = 0.00$). Small holders, due to lack of resources, highly preferred to adopt low cost practices such as crop rotation as compared to those of medium and large farmers. Medium and large vegetable growers showed significantly higher adoption of HYV seeds and use of farm credit as compared to small holders. The adoption rate of soil testing and use of green manure is also comparatively higher for medium and large vegetable growers than small-scale growers.

Post-Harvest Management Practices

Since agricultural products are not consumed immediately after harvest, post-harvest activities such as washing and cleaning, sorting and grading, dehydrating and drying, packing and storage play an important role in determining the final value of the product. Therefore, the production process of food and other agricultural products does not end when the crop is harvested. The income of a farmer does not depend only on total production but, to a large extent, it is determined by his/her ability to manage post-harvest losses and market the produce at optimum prices. Post-harvest spoilage and wastage of foodstuffs is a common phenomenon in both developed and developing countries.

Table 4.4 presents the chi-square statistics on adoption level of post-harvest practices across landholdings. Comparatively, the small holders found with lower adoption of post-harvest management practices (except sorting and grading) than those of counterparts. There is significantly higher adoption of washing and cleaning practices by medium and large growers than those of the small holders ($\chi^2 = 6.664, p = 0.007$).

Table 4.4: Chi-square statistics on adoption of post-harvest practices by farm size

Post-harvest management practices	Overall		Small holders		Medium & Large		Chi-square	Sig.
	N	%	N	%	N	%		
Washing and cleaning								
Adoption	158	28.5	110	25.8	48	37.5	6.664***	0.007
Non-adoption	397	71.5	317	74.2	80	62.5		
Sorting and Grading								
Adoption	198	35.7	157	36.8	41	32.0	0.963	0.191
Non-adoption	357	64.3	270	63.2	87	68.0		
Dehydrating and drying								
Adoption	131	45.0	97	43.5	34	50.0	0.890	0.210
Non-adoption	160	55.0	126	56.5	34	50.0		
Packaging								
Adoption	141	25.4	102	23.9	39	30.5	2.251*	0.084
Non-adoption	414	74.6	325	76.1	89	69.5		
Storage								
Adoption	185	33.3	136	31.9	49	38.3	1.833	0.107
Non-adoption	370	66.7	291	68.1	79	61.7		

***Significant at the 0.01 level, **Significant at the 0.05 level, *Significant at the 0.10 level.

Sales and Marketing Practices

Farmers need information and knowledge for minimizing post-harvest losses and selling their produce on remunerative prices. Similarly, they also require market related information to make appropriate marketing decisions (Shroeder, Ward, Mintert, and Peel, 1998). Decisions of farmers whether to sell the agricultural produce at the farm gate or travel to far away markets depends on many factors such as farm size, wealth level, level of surplus, connectivity and transportation facilities, proper market information, and reach to formal market institutions (Fafchamps and Hill, 2005; Shilpi and Umali-Deininger, 2007).

Table 4.5 provides chi-square statistics on adoption of sales and marketing practices by small holders. Small holders showed significantly higher adoption on selling the produce at organized market place such as *mandi* as compared to the medium and large farmers ($\chi^2 = 2.113$, $p = 0.088$). Whereas, the medium and large growers increasingly used market information and also resorted to marketing contract to minimize the market risk.

Table 4.5: Chi-square statistics on adoption of sales and marketing practices by farm size

Sales and marketing practices	*Overall*		*Small-holders*		*Medium & Large*		*Chi-square*	*Sig.*
	N	*%*	*N*	*%*	*N*	*%*		
Use of market information								
Adoption	97	17.5	68	15.9	29	22.7	3.094	0.054
Non-adoption	458	82.5	359	84.1	99	77.3		
Selling the produce at mandi								
Adoption	300	54.1	238	55.7	62	48.4	2.113*	0.088
Non-adoption	255	45.9	189	44.3	66	51.6		
Managing contract for selling the produce								
Adoption	44	7.9	28	6.6	16	12.5	4.764**	0.027
Non-adoption	511	92.1	399	93.4	112	87.5		

***Significant at the 0.01 level, **Significant at the 0.05 level, *Significant at the 0.10 level.

Factors Affecting Adoption of Agricultural Practices among Small-holders

Several studies have analyzed the factors that affect the adoption of innovative agricultural practices by agricultural producers (Feder and Slade 1984, Feder *et al.* 1985, Wozniak 1987, Agwu and Anyanwu 1996, Ramirez and Shultz 2000, Doss and Morris 2001, Park and Lohr 2005). The producers' characteristics such as age, education, income were investigated as important socio-demographic considerations that affect the innovation adoption across agricultural supply chain.

The results on coefficient estimate and significance levels for the parameters of the Poisson Count Regression Models on the adoption of innovative practices across the vegetable supply chain, viz., cultivation, post-harvest management and sales and marketing by the Small holders are presented in Table 4.5. Though, age and number of working family members are considered to be an important factor in the innovation adoption, the coefficient for these variables are not significant across the chain. Education and social category played significant role in the adoption of sales and marketing practices in vegetables. This implies that educated vegetable growers that belonging to socially lower group are inclined to adopt innovative cultivation practices for better income and employment. Income is a central socio-demographic factor that significantly affected the adoption of post-harvest practices. Overall, education and income are important factors that significantly influence the innovation adoption across the vegetable chain.

Farm characteristics are another set of variables influencing the adoption of innovative practices across the supply chain. Though, several empirical studies indicated strong positive relationship between farm size and adoption of farm technologies (Fedder *et al.*, 1985, Caswell *et al.*, 2001, Rahelizatovo and Gillespie, 2004), the result of this study indicated the overall operational landholding as a significant negative factor in adoption of innovative practices in vegetables. The findings indicate that vegetable growers with large landholdings are 8 per cent less likely to adopt the innovative practices as compared to small landholders. This is quite interesting to note that small vegetable growers are becoming more conscious in adopting modern agricultural practices for better productivity and profitability from their small farms. The vegetable growers who are practicing cultivation on leased-in land are 8 per cent more likely to adopt innovation in their cultivation practices.

Number of crops gown is another important farm related factor that significantly influenced the adoption of innovative cultivation practices and post-harvest practices. Area under irrigation, as expected, positively and significantly influenced the adoption of innovation in cultivation practices. Vegetable growers with more than 50 per cent area under irrigation are 20 per cent more likely to adopt innovation for effective vegetable cultivation. ICTs as crucial for modern agriculture particularly, have significant influence on the adoption of post-harvest practices and sales and marketing of vegetables. Overall, area under irrigation, number of crops grown and use of Information and Communication Technologies (ICTs) have significant influence on the adoption of innovative practices across the vegetable supply chain. Distance from market is also important factor that determines adoption of innovations across the vegetable chain.

When compared with expected impact of the independent variables on overall adoption of agricultural innovation, the socio-demographic variables such as education and income showed the similar result with significant and positive influence on the adoption. Age, social category and number of working members are resulted as non-significant factors. Likewise, the farm related variables such as landholding size and leased-in land are also resulted as non-significant factors. However, area under irrigation, number of crops grown and use of ICT have significant positive influence on the adoption, similar to the expected result. Distance from market is also resulted as non-significant factor.

Table 4.6: Parameter Estimates of Poisson Count Models

Parameters	*Adoption of cultivation practices*		*Adoption of post-harvest practices*		*Adoption of sales & marketing practices*		*Overall adoption of innovative practices*	
	β	*Predicted % change in adoption*	*β*	*Predicted % change in adoption*	*β*	*Predicted % change in adoption*	*β*	*Predicted % change in adoption*
(Intercept)	0.592**		–0.198		–0.644**		1.153***	
AGE	–0.002	–0.2	0.003	0.3	0.002	0.2	0.000	0.0
EDU	0.102	10.8	–0.029	–2.8	0.237**	26.7	0.083*	8.6
SOC	–0.055	–5.3	0.090	9.5	0.470***	60.0	0.061	6.2
INC	0.071	7.4	0.332***	39.4	0.019	1.9	0.145***	15.6
WRKMB	0.004	0.4	–0.012	–1.2	–0.029	–2.9	–0.005	–0.5
LANDSZ	–0.008	–0.8	–0.017	–1.7	0.013	1.3	–0.008	–0.8
LEASED	0.082	8.5	–0.066	–6.4	–0.049	–4.7	0.018	1.8
IRRG	0.188***	20.7	0.225	25.2	–0.015	–1.5	0.165***	18.0
CROPS	0.067***	6.9	0.034**	3.5	–0.003	–0.3	0.047***	4.8
ICT	0.008	0.8	0.079**	8.2	0.084*	8.7	0.041**	4.2
DISTANCE	–0.032	–3.1	0.085	8.8	0.019	1.9	0.013	1.3
Log Likelihood	–895.344		–800.791		–564.468		–1108.942	
Likelihood Ratio Chi-Square	60.379***	Df = 11	43.907***	Df = 11	27.494***	Df = 11	70.712***	Df = 11

***Significant at the 0.01 level, **Significant at the 0.05 level, *Significant at the 0.10 level.

Conclusions and Policy Implications

Small holders are the primary stakeholders who form bedrock for agricultural supply chain in India. Their adoption of technology and advanced agricultural practices can boost agricultural growth that may lead India to become food basket of world. The findings of the study indicate that the smallholders, due to lack of resources, preferred to adopt low cost technology and farm practices in their crop cultivation, post-harvest management and sales and marketing of produce. Small holders are highly involved in practicing crop rotation, sorting and grading, and selling the produce at organized market place such as *mandi*. In order to minimize agricultural risk, Small holders grew more crops as compared to the medium and large farmers. Lack of collateral is a major problem for Small holders to access institutional credit for entrepreneurial vegetable production.

The adoption of innovative agricultural practices is influenced by a small number of socio-demographic and farm-related variables. Regression analysis indicated that most socio-demographic factors did not influence adoption of post-harvest techniques except income level. Of 12 farm-related factors there were positive and significant coefficients for availability of irrigation, credit use, adoption of integrated pest management, adoption of new crops, diversification of farm activities, generation of marketed surplus, and market linkage through organized selling of vegetables.

These results can be used by policy makers, agribusiness companies and researchers involved in agribusiness systems, and stakeholders involved in the vegetable value chain. Agribusiness companies involved in manufacturing and retailing of vegetables and vegetable products may design their business/procurement plans as per the adoption behaviours of vegetable growers. Companies involved in generation of post-harvest technologies can better design their business strategies based on adoption by vegetable producers. Future research needs to identify new empirical models and additional explanatory variables to enhance predictability and improve understanding of farmers' adoption of post-harvest techniques.

Acknowledgement

This paper is a part of externally funded research project "Demand Assessment of Horticultural Commodities in Uttar Pradesh" sponsored by the Uttar Pradesh Council of Agricultural Research (UPCAR), Government of Uttar Pradesh, Lucknow.

REFERENCES

Agwu, A. E.; Ekwueme, J. N. and Anyanwu, A. C., (2008). "Adoption of improved agricultural technologies disseminated *via* radio farmer programme by farmers in Enugu State, Nigeria". *African Journal of Biotechnology*, 7, 1277-1286.

Birthal, P.; P. Joshi and A. Gulati. (2005). *Vertical Coordination in High Value Food Commodities: Implications for Smallholders,* IFPRI MTID Discussion Paper no. 85.

Busch, L. and C. Bain. (2004). "New! Improved? The transformation of the global agri-food system", *Rural Sociology*, 69(3). 321–346.

Cash, D.W., (2001). "In Order to Aid in Diffusing Useful and Practical Information Agricultural Extension and Boundary Organizations", *Science Technology Human Values,* 26, 431-453.

Caswell, M.; K. Fuglie; C. Ingram; S. Jans; C. Kascak. (2001). *Adoption of Agricultural Production Practices: Lessons Learned from the U.S. Department of Agriculture Area Studies Project.* Agriculture Economic Report-792, U.S. Department of Agriculture (USDA). Washington, DC.

Deshingkar, P.; U. Kulkarni; L. Rao; S. Rao, (2003). "Changing Food Systems in India: Response-sharing and Marketing Arrangements for Vegetable Production in Andhra Pradesh", *Development Policy Review*, 21 (5-6). 627-639.

Doss, C.R.; and M. L. Morris, (2001). "How does gender affect the adoption of agricultural innovations? The case of improved maize technology in Ghana". *Agricultural Economics,* 25, 27-39.

Feder, G., and R. Slade. (1984). "The Acquisition of Information and the Adoption of New Technology". *American Journal of Agricultural Economics*, 66, 312-320.

Feder, G.; R. E. Just; and D. Zilberman, (1985). "Adoption of Agricultural Innovations in Developing Countries: A Survey". *Economic Development and Cultural Change,* 33, 255-298.

Galloway, L., and R. Mochrie, (2005). "The use of ICT in rural firms: A policy-orientated literature review". *The Journal of Policy, Regulation and Strategy for Telecommunications*, 7, 33-46.

Gulati, A.; N. Minot; C. Delgado and S. Bora, (2007). "Growth in high-value agriculture in Asia and the emergence of vertical links with farmers". In: J.F.M. Swinnen, Editor, *Global Supply Chains, Standards and the Poor: How the Globalization of Food Systems and Standards*

Affects Rural Development and Poverty, CABI, Wallingford, UK (2007). 91–108.

Henson, S. and T. Reardon, (2005). "Private agri-food standards: Implications for food policy and the agri-food system", *Food Policy*, 30(3). 241-253.

Midmore, D. J. and H. G. P. Jansen, (2003). "Supplying vegetables to Asian cities: Is there a case for peri-urban production?" *Food Policy*, 28, (1). 13-27.

Park, T. A. and L. Lohr, (2001). "Organic pest management decisions: A systems approach to technology adoption". *Agricultural Economics*, 33, 467–478.

Rahelizatova, N. C., and J. M. Gillespie, (2004). "The Adoption of Best Management Practices by Louisiana Dairy Producers". *Journal of Agricultural and Applied Economics*, 36, 229-240.

Ramfrez, O. A., and S. D. Shultz, (2000). "Poisson Count Models to Explain the Adoption of Agricultural and Natural Resource Management Technologies by Small Farmers in Central American Countries". *Journal of Agricultural and Applied Economics,* 32, 21–33.

Rao, N. H., (2006). "A framework for implementing information and communication technologies in agricultural development in India". *Technological Forecasting and Social Change,* 74, 491-518.

Swinnen, J. F. M. and M. Maertens, (2007). "Globalization, privatization, and vertical coordination in food value chains in developing and transition countries", *Agricultural Economics*, 37(s1). 89-102.

Wozniak, G. D., (1987). "Human Capital, Information, and the Early Adoption of New Technology". *The Journal of Human Resources*, 22, 101-112.

5

Efficiency of Price Discovery in Agricultural Commodity Trading in India

Tanuj Nandan and *Niti Nandini Chatnani*

Introduction

Agriculture, along with its allied activities, contributes about 14 per cent of India's GDP at constant 2004-05 prices as per Central Statistical Organization (CSO) Advance Estimates, 2011-12. More significantly, it provides employment to close to sixty per cent of the nation's populace. In terms of both, arable and cultivated land area, India is one of the largest agrarian economies in the world. Yet, about 90 per cent of all farmers in the country fall in the category of marginal and small farmers, and contribute only 43 per cent of the total value of agricultural output. Further, India's agricultural sector has been characterized by low levels of informational and pricing efficiency, as is evidenced by the fact that only 25-30 per cent of the price paid by the consumer for agricultural produce is actually realized by the farmer, whereas in developed markets like the US, the farmer's share can be 60-65 per cent (Chatnani, 2008).

Revenues and incomes from agriculture can vary widely from year to year due to unforeseen weather, disease/pest infestations, and/or market conditions. When the unfavourable deviations in agriculture significantly reduce income in the short-term, there can be serious repercussions in the absence of effective risk management tools. Especially in case of developing countries that have remained commodity-dependent with a large proportion of dependent population it becomes crucial to manage backend and price risk (Cardinal Edge Management Services, 2008).

Organized Commodities Trading: Need and Genesis

Price risk can have significant implications for farmers, particularly those who fall under the category of small and marginal, and have less capacity to absorb losses. The hedging of price risk is possible through futures trading. As a regulated and efficient mechanism for trading of futures and options contracts based on commodities, any commodity exchange is established with the objective of providing buyers and sellers of commodities with a regulated mechanism for price discovery and price risk management. So, whether it is the Chicago Board of Trade (CBOT), the world's first commodity exchange established in 1848, or the Zhengzhou Commodity Exchange (ZCE), established in 1990, or the Multi Commodity Exchange (MCX) established in India in 2003, or the numerous other exchanges like them, it is clear that they have been intended to serve the genuine hedging needs of commodity producers and users.

CBOT, for instance, was created because the farmers who would bring their crops to sell to the Chicago sometimes had to burn or dump the crop into Lake Michigan because they could not sell it. In 1848, when the CBOT was organized, it was with the intention of giving farmers and other grain market participants the ability to manage the risk they faced on account of price volatility. This objective for setting up of commodity exchanges is today rather in contrast to the fact that most commodity exchanges are now a favourite platform for speculators who trade with the objective of gaining from the price volatility in commodity prices. While it is argued that speculators bring volatility to the prices of commodities on which contracts are traded on exchanges, it is these very speculators who allow the hedgers to transfer their risk.

Therefore, despite the arguments against the success of commodity exchanges in allowing for price risk management, since the 1980s most developing countries of the world have set up their own commodity exchanges, and these commodity exchanges have also come to occupy important places in their economies. It has been estimated that the volume traded on these exchanges are several times those traded in the physical markets or on the stock exchanges.

Development of Commodities Trading in India

In India too, commodity futures trading is almost as old as that in the United States, with India's first organized futures market, the Bombay Cotton Trade Association Ltd. established in 1875, followed by the setting up of Gujarati *Vyapari Mandali* in 1900 for futures trading in oilseeds, and many commodity exchanges across the country trading in jute, pepper, turmeric, potatoes, sugar, etc. However, consequential to the policy decisions taken after independence, mainly on account of the scarcity situation then prevailing, the development of commodity futures markets was negatively affected.

While in 1952, the Forward Contracts Regulation Act was passed with powers to regulate all transferable forward and futures contracts, the 1960s and 70s saw the Government of India suspend trading in several commodities like cotton, jute, edible oilseeds, etc. Agro Commodity markets in India did not develop their competitive strength partially due to government offering producers of most strategic agro commodities a zero-cost put in the form of a "Minimum Support Price" (MSP), and partially due to its monopoly in storage, transportation and distribution of agricultural produce. A ban on commodity futures trading only served to further weaken the agricultural commodity markets in the country.

For a country such as India, which has traditionally been an agrarian economy, strengthening the agricultural sector has been the key to addressing the challenges of declining contribution of agriculture to the country's GDP, food inflation and food insecurity, unemployment and rural poverty. Farmers face risks that can be avoidable, manageable or transferable. For the average Indian farmer, the risks arising from poor quality seeds and fertilizers are avoidable, and the risks of bad yield coming out of bad weather are manageable through insurance. Therefore, it has been critical to offer the farmers an opportunity to transfer to the risk-seeking speculators those risks that are neither avoidable nor manageable, and one such risk is price risk.

Instability of prices amounting to price risk has remained a major problem not just for the farmers, but also for the end-consumers and those in the marketing, trading and processing of agro commodities. This has warranted an examination of the entire agriculture value chain comprising production, processing, marketing, distribution and trade of agricultural products by policy makers. As price volatility in the agro-commodity sector has been a continuing challenge for Government machinery, it was being increasingly realized that Government support to commodity sector by way of continued market intervention or financial support by way of subsidy or minimum support price, etc., was neither sustainable nor desirable in the long run. It was, thus, agreed that only an effective market-based mechanism could sustain long in this front, motivating the Government to explore some alternative methodology to protect the commodity sector from price-volatility.

Exchange-based Commodities Trading in India

Government of India took the landmark decision to remove all the commodities from the restrictive list, and allowed setting up of new modern, demutualized Nationwide Multi-commodity Exchanges with investment support from public and private institutions (Kevin, 2010). While the ban on futures trading in select agricultural commodities was removed in the eighties, it was only in 2003 that commodity exchanges listing futures contracts were allowed to come into existence.

The Commodity Futures in India began its race from there with the emergence of National Multi Commodity Exchange (NMCE) of India Ltd., the first such exchange to be granted permanent recognition by the Govt. NMCE commenced futures trading in 24 commodities on 26th November, 2002 on a national scale and identifying the hedging needs of the commodity economy, the basket of commodities has grown substantially since then to include cash crops, foodgrains, plantations, spices, oilseeds, metals and bullion among others. And two other nationwide Commodity Exchanges, viz., Multi-Commodity Exchange of India (MCX) and National Commodity and Derivatives Exchange (NCDEX) emerged into Indian economy paving way to a healthy competition in supporting the Indian Commodity Sector.

As a professionally managed on-line multi-commodity exchange, NCDEX came up as the principal Agricultural Commodity Exchange in India and among the leading agricultural exchanges in the world. It was established on April 23, 2003, and commenced operations on December 15, 2003, with trading in a range of over 45 commodity contracts, both agricultural and non-agricultural. Today, NCDEX has over 830 members, and trading on the NCDEX platform is conducted in more than 35,000 terminals across 700 centres in India, with delivery of commodities possible through a network of over 775 delivery centres with a holding capacity of around 1.3 million tonnes.

The futures markets for agricultural commodities that have since come to be hosted on these commodity exchanges hold the key for strengthening of the agricultural sector, and have been identified as the best alternative and mechanism to help commodity sector, including the farmers, and protecting them from the adverse effects of price volatility. When NCDEX came into existence, it was expected to address many of the imperatives for effective risk management, which may be enumerated as:

- Creation of awareness and training,
- Access to market information,
- Institutional development,
- Formation of linkages with professional service providers and institutions,
- Adoption and trial of price hedging through commodity derivatives, and
- Validation and reconfiguration of existing commodity derivative contracts.

The Exchange has played a pioneering role in the online commodity markets in the country, implementing the concept of electronic delivery of commodities; dematerialized (demat) system of holding stocks, setting price benchmarks in spot markets, amongst others. NCDEX has been at the forefront of increasing participation in the commodity markets, conducting numerous awareness programmes across the country, in addition to disseminating the prices of commodities to the farmer community, enhancing their opportunities for better realization of their produce and bringing prosperity and economic freedom.

While offering trading in agricultural commodities like spices, cereals, oils and oilseeds, pulses, plantation products, fibres, etc., along with non-agricultural commodities like bullion, metals, and energy products, NCDEX distinguishes itself in being an exchange that offers offering futures trading in spices, as no futures on the flavouring commodities are traded on any of the leading international exchanges. NCDEX's spice basket consists of Chilli (*Lal Mirch*), Coriander (*Dhaniya*), Cumin (*Jeera*), Black Pepper (Kali Mirch) and Turmeric (*Haldi*).

Chilli Market

Given below is a quick fact sheet about the most common spice from this spice basket, and the second largest traded spice in the world—chilli:

1. The chilli crop takes 3-4 months to be ready, with two crops of chilli produced each year in India, one during the dry season from April to August, and another during the wet season from August to December..
2. There are several varieties of chilli grown in India, and the most popular ones are *Sannam*, LCA 334, *Byadgi,* Wonder Hot, *Jwala,* etc.
3. India is the world's largest producer, consumer and exporter of chilli. While global production of chilli stands at about 20-25 lakh MT per annum, India contributes about 10-12 lakh MT of it, followed by China, Nigeria, Peru, Bangladesh, Hungary, Spain, Mexico, Pakistan, Moroco and Turkey. Major importers of Indian chilli are Sri Lanka, Bangladesh, United Kingdom, Germany, France and United States of America.
4. Around 90 per cent of India's production is consumed domestically, while 25 per cent of the global chilli exports come from India. Chilli , with the first position in spices exported from India, constitutes nearly a third of total spice export. The chilli export from India takes the form of dried chilli, chilli powder, pickled chilli and chilli oleoresin
5. Guntur, in Andhra Pradesh, is Asia's largest market for chilli, and about 80 lakh to 100 lakh bags of chilli, weighing 35-50 kgs each are traded during the season in the Guntur market alone. The marketing season takes off in the first week of February, peaks during April, and closes by mid of May. The area under chilli cultivation in Andhra Pradesh is 26 per cent of the total under chilli in India, while 50 per cent of the production comes from Andhra Pradesh. Of this, Guntur accounts for 38 per cent of the production, followed by Khammam, Warangal and Prakasam. And domestically, while Andhra Pradesh contributes 50 per cent of the total production of chilli, other States also contributing to chilli production are Karnataka, Orissa, Maharashtra, West Bengal, Rajasthan and Tamil Nadu.

The major centres for chilli trading in India are:

- *Andhra Pardesh*: Guntur, Warangal, Hyderabad, Cuddapah, Vijayawada.
- *Karnataka*: Dharwad, Mysore, Hassan, Bengaluru, Bellary, Ranibennur, Hubli, Byadgi.

- *Maharashtra*: Nasik, Ahmednagar, Sholapur, Aurangabad, Nanded, Amravathi, Lasalgaon.
- *Tamil Nadu*: Pollachi, Ramanad, Madurai, Trichi, Thnai, Dindigul, Virudunagar.

An examination or the value chain for chilli shows the following scheme:

Farmer – Commission Agent (*Kachcha Arhatiya*) – Trader (*Pakka Arhatiya*)

Processor – Wholesaler – Retailer – End Consumer
Exporters —

Exchange-based Trading of Chilli

Exchange trading in chilli was started towards the end of 2005. The unit of trading of the NCDEX Chilli Futures Contract is 5 MT, and the basis price is the price of chilli (*Paala*) traded as LCA 334 ex-warehouse Guntur, inclusive of all taxes and market cess. The exchange also maintains its accredited warehouses for chilli receipt and delivery at locations in Guntur and Warangal.

Since the NCDEX Chilli Futures Contract is benchmarked to the Guntur Chilli *Mandi*/Yard, it would be of help to understand that the Guntur Chilli Yard is owned and regulated by the Agricultural Market Committee, Guntur. This yard is located at a distance of 5 km from the main Guntur town, and spreads across nearly 50 acres between National Highway and Nagarjuna Sagar State Highway. Reputed as Asia's largest chilli market, it attracts produce from different parts of the State, and buyers from all over the world, generating the highest market fee among all yards in the State.

The yard has continued to rely on the tried and tested method of selling agricultural product through the commission agent who also finances the farmer carrying out chilli production. Chilli brought to the yard is sold in the open auction method through 575 commission agents to 375 chilli purchasers and exporters. The sale of chilli in the yard starts with a model auction at 7.00 am, and with sale prices announced through a public announcement system, traders procure chilli after examining and verifying samples. Trading closes by 2.00 p.m., and the farmer receives spot payment.

Yet, the impact of the emergence of Chilli Futures Contracts on

NCDEX can be better understood by understanding the impact of this development of the participants at various levels of the supply chain. A research study was carried out for this purpose, and the methodology involved unstructured interviews with participants like farmers, commission agents, traders, processor and exporters in the Guntur market, with presence in the chilli trade of more than 15 years. This requirement of experience was placed to select only those participants who have trade in the chilli markets before the emergence of the commodity exchanges and the launching for futures contracts on chilli.

Study on Efficiency of Price Discovery

Through this chapter, the authors attempt to establish that in the Indian context, the introduction of futures trading has resulted in an efficient price-discovery process. The introduction of exchange-based electronic trading platforms has facilitated communication of price-related information to traders across the world. At the same time, the ability of entering into trades and offsetting them without the physical interchange of commodities has brought in price-risk hedgers and arbitrageurs, who work towards minimization of price differentials, and thereby lead to price stability. Further, the fact that this market allows the participation of a large number of traders with market intelligence and price judgment, ultimately results in efficient price discovery.

The Research Question

Do participants across the value chain of the chilli trade consider the price discovery mechanism on commodity exchanges in India to be efficient? (Pricing efficiency may be defined as the absence of significant difference between the price prevailing on the exchange, and a 'fair' price)

Research Methodology

- Commodity chosen: Chilli
  Justification: India is the world's largest producer, consumer and exporter of chilli in the world.
- *Exchange*: NCDEX is an exchange that offers futures trading in spices, as no futures on the flavouring commodities are traded on any of the leading international exchanges.
- *Mandi* (Market): Guntur Chilli *Mandi*/Yard (Andhra Pradesh).

- The NCDEX Chilli Futures Contract is benchmarked to the Guntur Chilli *Mandi*/Yard. Guntur is Asia's largest market for chilli.
- *Mode of Data Collection*: Unstructured Interview with Participants.
- *Filter*: Participants in the study are required to have a presence of more than 15 years in the chilli trade.
- *Total sample size*: 120.

Findings

Some of the farmers were aware about the existence of futures in chilli on commodity exchanges, but had shown no interest in participating on selling on the exchange platform, and they cited five main reasons for this:

(1) Absence of the volumes required to trade in a contract of the exchange.
(2) The need to have a count to participate in exchange-based trade.
(3) The requirement of margin money and the hassle of being marked-to-market.
(4) The fear of being cheated by the broker; while having complete faith on the commission agent.
(5) Satisfaction with the operations of the local yard.

Interaction with the commission agents also revealed a similar story, i.e., the coming of commodity exchanges and contracts on chilli has not had any significant impact on their traditional role in the chilli value chain. Demand and supply in the market is the main price determinant, and the commission agents continue to earn a percentage of the sale value from farmer to trader. However, an interaction with the traders as well as the exporters provided some valuable insights into the impact of the exchanges on their ways of participating in the trade.

Most of the traders, being fully aware of the exchanges and the benefits of trading on them, hold accounts with brokers like Religare Commodities, India Infoline and SMC Commodities. They receive end-of-day spot and futures prices from the exchanges *via* messages in mobiles. In fact, most of these traders subscribe to the mobile messages updates from commodity trading firms, and use this information in their trading decisions. However, their usage of the commodity exchange platform is limited to times of extreme price volatility, typically:

(a) When the price of chilli is very low compared to *mandi* price, the trader/exporter will buy chilli on the exchange and take delivery form the designated warehouse.
(b) When they need to quote prices to their customers, the trader/exporter will use the price signals form the commodity exchange.

When asked to specify why they trade on the exchanges only in exceptional situations, the responses that came were similar to those received from the farmers:

(1) Trading chilli on the exchange involves having an account with a broker, and in the case of these traders, and allowing their brokers to place orders for them, as the traders have low levels of awareness and/or literacy. This reliance on the broker is discomforting for the trader, who likes to have complete ownership over his trades.
(2) The requirement to have quality of chilli tested by the assayer at the exchange accredited warehouse, and the cost and effort involved in delivering to or receiving from the warehouse is an inconvenience to the traders.

However, all participants interviewed during the research, namely, the farmers, the commission agents, and the traders, had confidence on the prices discovered and reported from the exchange. Therefore, the efforts over the last two years by exchanges jointly with Forward Markets Commission to install Electronic Price Display Boards at *mandis* and *Kisan Vikas Kendras* (KVKs) in various States, as a part of the Price Dissemination Project, have been valuable. The Price ticker boards display both Spot and Future prices of various commodities, getting its on-line Data feed from all the Commodity Future exchanges and AGMARKNET, with the potential to help farmers to take informed decisions on crop selection and marketing strategies.

Conclusions

The physical markets for chilli are well developed, and even though the exchange actively disseminates the prices of chilli across India, chilli trading on the exchange platform has not gained the participation of chilli farmers and commission agents. Some of the reasons that can be identified for this are:

1. An established chilli market for trading chilli being already in place, this traditional trading system has earned the confidence of farmers and continues to dominate as their market of choice.
2. Awareness and assurance required for participating on the exchanges among the farmers, commission agents and traders is still low, and the incentive to make an effort is also low as the physical *mandi* or yard system that brings together farmers, commission agents and traders on a single platform is efficient and popular among all participants.
3. Exchange trading is considered cumbersome as it involves quality and storage issues, and requires significant amounts of paperwork.
4. The need to rely on brokers as intermediaries for trading on the exchange is not liked by the traders, and they rely on the exchange only for price information.

Acknowledgement

The authors are indebted to Mr. Anil Kumar Kolla, student of IIFT's MBA (IB) 2011-12 Batch, for carrying out the research at the Guntur Chilli Yard.

REFERENCES

Cardinal Edge Management Services (2008). Enabling Farmers to Leverage Commodity Exchanges. Report submitted to the Multi Commodity Exchange of India.

Chatnani, N.N. (2008). Commodity Futures Markets in India: Rationale and Relevance for Farmers. ICFAI Analyst.

Kevin, S. (2010). *Commodity and Financial Derivatives*. New Delhi: PHI Learning Private Ltd.

http://agmarknet.in (Accessed on 05/Mar/2012)

http://fmc.gov.in (Accessed on 02/ Mar/2012)

http://www.indianspices.com (Accessed on 02/ Mar/2012)

Mohanty, S. (2008). Risk in Indian Agriculture, in Prasad B.V.S. and Mohanty S. (eds.) *Agricultural Risks Management*. Hyderabad: ICFAI Press.

http://www.ncdex.com (Accessed on 05/ Mar/2012).

Report of the Expert Committee to Study the Impact of Futures Trading on Agricultural Commodity Prices (2008). Report submitted to the Ministry of Consumer Affairs, Food & Public Distribution Government of India.

Sanders, D.R. and M. R. Manfredo, (2004). Reconsidering the Necessary Condition for Futures Market Efficiency: An Application to Dairy Futures. Paper presented at the NCR-134 Conference on Applied Commodity Price Analysis, Forecasting, and Market Risk Management, St. Louis, Missouri, April 19-20, 2004.

Sporleder, T. L. (1989). Pricing Efficiency in Agricultural Markets: Discussion. *Western Journal of Agricultural Economics*, 14(1): 122-125.

Section III
Small Farmers and Viability

6

Strategy for Sustainable Development of Small Farmers

Amarendra Pratap Singh and *R.P. Singh*

Introduction

Land is the main source of agricultural production and provides livelihood security for 65 per cent of the people. Since the Green Revolution, the country has produced enough food to feed itself, but it has not yet been able to wipe out mass hunger. Currently, 40 per cent of the population is malnourished – a decline of only 10 per cent in the past three decades. Following a series of neoliberal economic reforms in 1991, India's GDP has doubled, but despite this, about 53 million more people now go to bed hungry every night. Food prices have recently increased adversely affecting small farmers and poor families, who spend more than 60 per cent of their incomes on food. Unfortunately, small-scale producers have not benefited from high retail prices for food either. The country is in the midst of both an agrarian crisis and a nutrition crisis.

Food as well as nutritional security of the country very much depends upon the economic viability of small farmers. Unless their interest is protected through remunerative prices of their produce and increasing cost of inputs are checked they will not actively participate in farming. Evidences indicate that most of the farmers are not interested in farming occupation. For example, more than one-third farmers are engaged in farming due to compulsion rather than choice. The main reasons for dislike of farming occupation mentioned by the farmers are that it is not profitable

(27%), risky (8%), and lack social status (2%). Moreover, the level of dissatisfaction varies across States. More farmers are dissatisfied in poor States namely, Bihar (51%); Orissa (47%); Jharkhand (47%); Chattisgarh (46%); and Uttar Pradesh (46%) than in Southern States such as Tamil Nadu (31%) and Andhra Pradesh (24%). While more than 40 per cent of farming households feel that given a choice they would take up some other career. A large number of farmers are committing suicides even in agriculturally prosperous areas. This is a big challenge before the country to retain them in farming or provide alternate source of employment and income.

Land is highly unequally distributed and a large part of prime agricultural land, near cities and small towns, is being diverted to non-productive use. In fact cultivated area is constantly declining while population is still increasing at a faster rate putting increased pressure on limited cultivated area for increasing food production. Land can be successfully cultivated only if it is economically viable. No one can go on cultivating land unless he/she gets an adequate return from his land. However, sustainability of farming system requires that it is economically viable, technically feasible, environmentally sound, socially acceptable, administratively manageable and politically desirable. Strategy for accelerated inclusive growth must be agriculture-led economic growth and farmer-centric. It is to be noted that only alternative to feed India is improving the productivity and protecting the interest of large number of small farmers. Country cannot depend upon the other countries for food security.

Under these circumstances, the question arises that who will provide the nutritious food the growing population of this country and how these large number of small farmers can be empowered to accept the challenges and opportunities of new developments? What socio-economic policies shall facilitate the empowerment?

Structural Changes in Landholdings in India

Between 1970-71 and 2005-06, the total number of operational holdings in India increased from 71.01 million to 128.89 million and area under operational holdings declined from 162.18 million ha to 156.62 million ha. This resulted in the reduction of average farm size from 2.28 ha to 1.21 ha. In the same period, the share of small and marginal holdings in operated area doubled. Small holders now cultivate 42 per cent of operated land and constitute 84 per cent of total landholdings. Moreover, the small farmers use land more intensively and higher irrigation compared to large farmers (Table 6.1). Growth in rural population and subdivision of families are the main reason for increase in number of smaller holdings in India. It is interesting to note that since 1970-71, both the number of landholdings and rural population increased exactly at the same rate (1.76%). This broadly reflects the outcome of inheritance pattern prevalent in India.

Table 6.1: Comparative Picture of Small and Large Farmers

Indicators	*Marginal Farmers (Less than 1 ha)*	*Small Farmers (1-2 ha)*	*Other farm size*
Share in number of land holdings (%)			
1970-71	51	19	30
1995-96	62	19	19
2005-06	65	19	16
Share in operational Holdings (%)			
1970-71	9	12	79
1995-96	17	19	64
2005-06	21	21	58
Cropping Intensity (%)			
1981-82	134	128	124
1991-92	137	130	126
2001-02	139	128	128
Extent of area under irrigation (%)			
1980-81	40	33	27
1990-91	44	36	33
2000-01	51	39	39

Source: Department of Agriculture and Cooperation, Agricultural Census Division.

Small farmers, accounting for India's 83 per cent of all farmers, 45 per cent of the cultivated land, and 50 per cent of the total agricultural production provide food security for nearly 1 billion people. Development of small farmers is vital not only for Indian agriculture and agrarian economy, but also for achieving the Millennium Development Goals for reducing hunger and poverty in the world as a whole. More than half of the Small holders possess less than 0.5 ha land, referred to as sub-marginal farmers, and their number has been increasing with the increase in the population. Thus, the small is getting smaller. Despite their higher per unit productivity, the extremely small and fragmenting holdings are becoming economically nonviable, and the sub-marginal and marginal farmers are swelling the ranks of the hungry and poor.

Farm Size Productivity

The relationship between farm size and productivity has been intensely debated in India. A large number of studies during the 1960s and 1970s provided convincing evidence that crop productivity per unit of land declined with an increase in farm size (Sen, 1962, 1964; Mazumdar, 1965; Khusro, 1968; Hanumantha Rao, 1966; Saini, 1971; Bardhan, 1973; Berry, 1972). This provided strong support for land reforms, land ceiling and various other policies to support Small holders on ground of efficiency and growth. Subsequently, various analysts started exploring reasons or factors for higher productivity of Small holders (Berry and Cline 1979; Bhalla, 1979; Binswanger and Rosenzweig 1986; Dong and Dow 1993; Frisvold 1994; Raghbendra *et al.* 2000) and some of them even questioned the inverse relationship between farm size and productivity. Bhalla and Roy (1988) observed that the inverse relation between farm size and productivity weakened and disappeared when soil quality variable was included in their study. Chadha (1978) analyzing farm level data for three agro-climatic regions in Punjab for 1969-70, reported that the inverse relationship had ceased to hold in more dynamic zones.

Ghose (1979) argued that an essential precondition for the existence of the inverse relationship phenomenon is technical backwardness implying that with the advances in technology the inverse relationship will vanish. Similar to this, Deolalikar (1981) observed that the inverse size-productivity relationship cannot be rejected at low levels of agricultural technology in India, but can be rejected at higher levels. Rudra (1968) concluded that "there is no scope for propounding a general law regarding farm size and productivity relationship". Chattopadhyay and Sengupta (1997) in the context of West Bengal, reported that the inverse relation between farm size and productivity was stronger in agriculturally developed regions. On the other hand, Hanumantha Rao (1975) and Subbarao (1982) reported a positive relationship between farm size and productivity and attributed this to higher application of fertilizer and other cash-intensive inputs on large farms. Dyer (1997) argued that the inverse relationship is neither a product of superior efficiency on the part of small farms nor is it due to better quality land on the small farms but arises from the desperate struggle for poor peasants for survival on below subsistence plots of land. Hence, Dyer (1997) opined that redistribution of land on the basis of the inverse relation argument, far from alleviating poverty and creating employment opportunities, will only deepen and perpetuate extreme levels of exploitation and poverty.

Foster and Rosenzweig (2010) using plot level panel data (over the span 1999-2008), of the Rural Economic Development Survey (REDS) of the National Council of Applied Economic Research (NCAER), and using a model incorporating supervision costs, risks, credit-market imperfections and scale economies associated with mechanization, report that small-scale farming is inefficient in India. Thapa and Gaiha (2011) using all India survey, REDS, 2006 of NCAER data, analyzed the farm size crop yields relationship, using the Kernel density function, and observed that the relation varies with food commodity group. They also report that "while much lower fractions of Small holders are concentrated in lower ranges of yields compared with medium- and large-landholders, segments of Small holders also obtain very low yields".

Fan and Connie (2005) show that to increase labour productivity, and, therefore, farmer's income, either land productivity has to increase or land to labour ratio has to improve. Given the consensus that smaller farms have a lower land-labour ratio than large farms, Dyer (1997) and Havnevik and Skarstein (1997) argue that smaller farms enjoy higher land productivity in the short term, but over the long-term land productivity tends to drop. They argue that this long-term drop in land productivity results from over intensive cultivation of the land in order to maintain labour productivity, when more and more people need to survive on the same small area of farmland, and as the smaller farms are resource-poor to invest in preserving soil fertility, soil productivity eventually becomes exhausted and land productivity drops. A similar reasoning is given by Hazell (2011) who also maintains that many of the advantages of Small holders disappear as countries develop. The reasoning given for this is that as the per capita income rises, the economy diversifies and workers leave agriculture and the wage rate goes up. It then becomes more efficient to have progressively larger and more mechanized farms. Peter Rosset of Institute for Food and Development Policy indicates that smaller farms are 200 to 1,000 per cent more productive per unit area. Smaller farms with the more complex farming system get more total production per unit of cultivated area because they use more of the available niche space; tend to invest more labour in their land.

However, considering the growing number of small farmers, it must be noted that a food secure and peaceful India is in the hands of her small farmers. Without them, India will be a food insecure, violent and undemocratic society. It has been also observed that output per acre of investment is higher on small farms than on large farms. Thus, if a crowded, capital-scarce country like India has a choice between a single

100 acres farm and forty 2.5 acres farms, the capital cost to the national economy will be less if a country chooses the small farms (Former Prime Minister of India, Choudhary Charan Singh).

Status of Small Farmers

Indian agriculture is dominated by large number of small farm operators, which is likely to increase further due to subdivision of families but will continue to play a very important role in ensuring the food and nutritional security to the ever increasing population. Majority of the small farms cultivate mainly low value, subsistence crops. In the absence of adequate farm and non-farm employment opportunities, they are also forced to live below poverty line. The situation is likely to worsen because of the growing pressure of population on land and the limited scope of increasing additional production through subsistence farming. However, these small holdings are not homogeneous and vary in terms of not only size but resource base. There are small farm owner as well as small farm operators and small farm producers. The small farmers are also better stewards of natural resources which offer much more sustainability in the long run. The results from the nationwide surveys of farm households clearly show that the smaller size of holdings in India have been using higher doses of inputs, making more intensive use of land and adopting new technology on a much larger scale compared to farms in the larger size categories.

Economic Viability of Small Farmers

Planning Commission norm of poverty line for rural areas shows that a farmer operating less than 0.64 ha area will be under poverty. While based on the Tendulkar Committee norm, a minimum 0.8 ha of land area is needed to keep a farm family above the poverty line, if this family lives only on agricultural income. This implies that 62 per cent of farmers in India, who own less than 0.80 ha of cultivable land, would be under poverty if they do not have an opportunity to earn income outside agriculture. This figure is going to rise further with the ongoing division of landholdings (Chand, Ramesh, 2011).

While the small farm in India is superior in terms of production performance, it is weak in terms of generating adequate income and sustaining livelihood. Tiny holdings below 0.8 ha do not generate enough income to keep a farm family out of poverty despite high productivity. Nearly three-fourths of small farmers in India fall under poverty if they do not get income from non-farm sources. (Ramesh, Chand, 2011). It is thus clear that with the present level of productivity, three-fourths of the population of Small holders cannot meet their livelihood from farm income alone. There are mainly two ways to improve their income and livelihoods (1) increase in land-man ratio, which is possible only if a sizeable segment of small land operators is moved out of agriculture; and (2) provide alternative sources of employment to Small holders in or around their habitation to supplement their farm income (Table 6.2).

Table 6.2: Extent of Irrigation and Fertiliser Use by Small Farmers

Indicators	*Marginal Farmers (Less than 1 ha)*	*Small Farmers (1-2 ha)*	*Other farm size*
Fertilizer consumption per ha of net sown area (kgs)			
(a) Total Sown Area			
1981-82	55	48	40
1991-92	99	85	76
2001-02	175	129	119
(b) Irrigated Condition			
1981-82	100	98	99
1991-92	145	140	146
2001-02	252	206	203
(c) Unirrigated Condition			
1981-82	24	22	16
1991-92	58	48	38
2001-02	96	76	61

Source: Department of Agriculture and Cooperation, Agricultural Census Division Constraints faced by the Small Farmers.

In India, almost all small farms practice multi-diversified farming and grow a number of crops even on small acreage and fragmented plots. Many of them undertake crop and livestock farming in a symbiotic manner. But this kind of farming does not necessarily yield enough returns for the sustenance and upward movement of small farm families. Therefore, what is important for real diversification is that small farmers do not just grow a large number of less productive, subsistence crops and undertake traditional livestock farming. Instead, they should cultivate a few selected high yielding, high income generating and eco-friendly crops and livestock enterprises which fit well into the agro-climatic environment of each region. In general, these small farmers face several constraints such as physical constraints; socio-Economic Constraints; technological constraints; infrastructural constraints; and institutional constraints (Haque, 1985).

Physical Constraints

The major constraint for efficient use of land is subdivision and fragmentation of holdings as these farms get deprived of the necessary economies of scale. Small and marginal farmers lack ability to invest in crop and non-crop enterprises which require high investment of capital and labour inputs. They also find it difficult to spend time to learn about, undertake and assume risks of adopting any new agricultural economic activity, unless the relative economic gains from such activity have already been demonstrated elsewhere. In general, markets and price information do not exist at all for some of the high value, perishable commodities which constrain the process of diversification.

Technological Constraints

The present level of technological development for some of the high value fruits and vegetable crops is not only inadequate, but the available technology does not suit various agro-climatic situations. A number of micro level studies have pointed out that non-availability of good planting material and high incidence of pests and insects act as major constraints to the adoption of some high value crops. Therefore, generation of appropriate technologies including low cost biotechnologies and their quick dissemination would be essential for sustainable growth of smallholder agriculture.

Infrastructural Constraints

Since infrastructure development is highly concentrated in certain developed regions. The small farms in backward regions do not have access to basic infrastructural facilities of road, transportation, linked market, storage facilities, cold storage, irrigation and power, etc.

Institutional Constraints

Evidence suggest that effective and functional institutional arrangements at the grass root level are important for providing necessary credit, capital and marketing facilities. In this regard, it is necessary to further strengthen the cooperatives, *Panchayati Raj* system, and existing network of *Krishi Vigyan Kendras* or extension workers. Moreover, government alone may not be able to provide all such requirements, hence private market entrepreneurship or a system of contract farming can also be encouraged to participate in the development process of small farmers.

Credit Constraint

Most of the farmers, especially small farmer face serious constraints of finance to meet the requirement of required inputs. Hence, easy access to credit at low rate of interest well in time is necessary for them to make investment in farming. Most of these farmers do not have access to institutional credit and are dependent on informal sources such as private money lenders, friends and relatives. A large proportion of them are highly indebted and in the absence of adequate income they remain in debt for generations. It is clear from Table 6.3 that still nearly half of the rural households are indebted in India. However, this figure varies across States as in Uttar Pradesh more than 82 per cent of farmers are indebted than only 18 to 20 per cent in Jharkhand and Assam. (Table 6.3 and Annexure 6.2).

Table 6.3: Proportion of Indebted Rural Households in Different States

Sl. No.	*Range of indebtedness (%)*	*Name of the States*
1.	Above 75 %	Andhra Pradesh, Tamil Nadu.
2.	60 to 75 %	Punjab, Karnataka, and Kerala.
3.	50 to 60 %	Maharashtra, Haryana, Rajasthan, Gujarat, Madhya Pradesh, and West Bengal.
4.	40 to 50%	Orissa, Chhattisgarh, and Uttar Pradesh.
5.	20 to 40%	Himachal Pradesh, Jharkhand, Jammu & Kashmir, and Bihar.
6.	Less than 20%	Assam and Uttrakhand.

Source: Report No. 498(59/33/1), Situation Assessment Survey of Farmers: Indebtedness of Farmer Households, National Sample Survey 59th Round (January-December 2003).

Moreover, if we compare the extent of indebtedness across various farm size classes it becomes clear that even large farmers are equally or marginally more indebted than marginal and small farmers. But, most striking feature is that extent of loan on large farmers (operating land above 10 hectares) are considerably high (almost 6 times) compared to farmer operating land holding of 1-2 ha. Moreover, even the institutional credit facilities are monopolized by a few relatively developed regions (Table 6.4).

Table 6.4: Indebtedness of Farmer Households in Different Size Classes

Land Possessed (hectare)	*% share of total farmers households in each Class*	*% share of indebted farmer households in each Class*	*% of Indebted Farmer Households to Total*	*Average Outstanding Loan Amount (Rs)*
<0.01	1.40	1.30	45.3	6121
0.01-0.40	32.80	30.00	44.4	6545
0.41-1.00	31.70	29.80	45.6	8623
1.01-2.00	18.00	18.80	51.0	13762
2.01-4.00	10.50	12.50	58.2	23456
4.01-10.00	4.80	6.40	65.1	42532
>10.00	0.80	1.20	66.4	76232
All	**893.50 (100.00)**	**434.24 (100.0)**	**48.6**	**12585**

Source: Report No. 498(59/33/1), Situation Assessment Survey of Farmers: Indebtedness of Farmer Households, National Sample Survey 59th Round (January-December 2003).

Government has distributed a large number of Kisan Credit Cards to facilitate farmers to have access to institutional credit, but nearly 23 per cent of total Kisan Credit Cards are issued to three major States namely Uttar Pradesh, Andhra Pradesh, Maharashtra only. However, most of credit cards have been distributed by cooperative banks (48.2%) followed by commercial banks (39.5%) and rest about 12 per cent by Regional Rural Banks (Table 6.4. and Annexure 6.3). Existing unequal infrastructural facilities might further accentuate regional disparities in development.

Scope for Improvement

Moreover, it is true that small farmers use the available resources with them more intensively and judiciously. But, since the technology is neutral to scale it is the resource base and access to support services such as credit, extension, availability of quality inputs, etc., substantially affect the productivity of small farmers. The small farmers tend to diversify their farming to meet the subsistence requirement and also to provide food to other people. They have tremendous scope for increasing productivity and can intensify biodiversity through conservation and rejuvenation resulting into higher biological productivity and stability and sustainability of agriculture.

The small bio-diverse farms have higher productivity than large industrial farms. Higher biological productivity translates into higher incomes for small farmers. Results of various studies indicate that in Rajasthan, monocultures of Pearl Millet gave Rs. 2,480 of net profit per acre, whereas a bio-diverse farm of Pearl Millet, Moth Bean, Sesame gave Rs. 12,045, a difference of nearly Rs. 10,000 per acre. In Uttarakhand, a monoculture of paddy gave Rs. 6,720 per acre, whereas a bio-diverse farm gave Rs. 24,600 per acre, a difference of Rs. 16,000. In Sikkim, a monoculture farm of maize gave Rs. 4,950 per acre while a mixed farm of maize, radish, *Lahi saag* and peas gave Rs. 11,700. Navdanya's rice and wheat farmers have doubled the production of rice and wheat by using indigenous seeds and organic methods. Jhumba rice in Uttarakhand has 176 quintal per ha of biomass production compared to 96 quintal per ha of *Kasturi*, a high yielding rice variety. The paddy yields are 104 and 56 quintal per ha respectively. Farmers in West Uttar Pradesh have got 62.5 quintal per ha using a native wheat variety 308 for organic production compared to 50 quintal per ha for chemically produced wheat (Vandana Siva, 2007).

In contrast, large farmers and industrial farming has serious limitations on increasing agricultural productivity. These large farms exploit and deplete natural resources reducing the soil fertility through indiscriminate use of chemical fertilizers and use of heavy machinery. These large farmers use excessive water as chemical farming needs ten times more water than ecological farming. They intensify external inputs such as agrichemicals and fossil fuels, which lower the productivity, and lead to non-sustainability and economic and ecological vulnerability.

Small bio-diverse farms based on internal inputs are, in fact, the only promise for increasing agricultural productivity, whether productivity is defined in terms of biological productivity or in terms of financial returns, or in terms of energy. Large industrial farms use ten times more energy than they produce as food, most of the energy goes to pollute the atmosphere and destabilize the climate. Moreover the industrial model of high external inputs imposed on small farmers leads to increased indebtedness and suicides.

Small Farmers and Multinationals

Under WTO regime, allowing entry of multinationals and corporatisation of agriculture has far reaching consequences for food security and livelihood of small farmers in India. Food supply increasingly falling under the control of giant multinational corporations and global large agribusiness corporate houses significantly affect the global agricultural economy, using "free trade" agreements to turn food into a commodity for profit rather than a human right. This has disastrous effects on farmers, food security, and the environment.

Evidence from other developed countries also indicates that various trade agreements and dominance of multinationals have adversely affected the small land holders. For example, implementation of the North American Free Trade Agreement (NAFTA) in 1994, 'liberalized' trade between Canada, the U.S. and Mexico, substantially affected the farmers' income in all three countries. Millions of small farmers lost their land while agribusiness corporations reaped huge profits. It is not unexpected that globalization of agriculture would put transnational corporations in control of the entire food supply and market forces, rather than national policies set by democratically elected officials, would control agricultural food systems. Under this scenario, each country would only produce a few export commodities, wiping out local food production, small family farms, and greatly compromising global food security. As a result, the human right to food would be dependent on multinational corporations and markets, increasing the risk of hunger and famine worldwide. For promoting local economic development one must realize that local food production is key to food sovereignty. Hence, there is a need to have entirely different approach to agriculture and trade, which gives food sovereignty, security, and the preservation of rural livelihoods.

Worldwide, the trade in fresh vegetables and fruit is increasingly controlled by a small number of multinationals and Super Markets. The fact that free trade in agriculture and services has a negative impact on small farmers and producers in developing countries is being neglected in the WTO negotiations. Both in the global North and South, Super Markets own a growing part of the market. A mere 2 per cent of farms in the United States control more than half of American agricultural sales. Over 73 per cent of the nation's farms share less than 7 per cent of the market value of agricultural products, while 7.2 per cent of farms receive 72 per cent of the market value of products sold. Eighty-two per cent of U.S. corn exports are controlled by three agribusiness firms—Cargill, Archer Daniels Midland (ADM), and Zen Noh. While family farmer incomes have plummeted during the first 7 years of NAFTA, ADM's profits went from $110 million to $301 million, while ConAgra's grew from $143 million to $413 million.

In the Netherlands, more than 80 per cent of all fresh fruit and vegetables are sold at Super Markets, of which the 5 largest Super Markets own more than 50 per cent of the market. The increasing concentration of Super Markets has an effect deep into the supply chain as Super Markets are increasingly able to stipulate who produces what, where and for what prices. The consequences for small farmers are significant; they cannot meet the requirements of the Super Markets, who want products for a low price, of the best quality, in large quantities and throughout the whole year.

The domination of the financially strong foreign multinational supermarket chains is a threat for the local producers who cannot meet the new requirements. As a necessary element to human survival, food is a human right. Small, local family farms are the bedrock of traditional rural communities and global food security—the ability of countries to produce the food they need to survive. Yet the global food supply is increasingly falling under the control of giant multinational corporations.

In reality, globalization of food production represents an unprecedented hijacking of the global food supply for corporate profit. The drive toward globalization of agriculture would put transnational corporations and markets, increasing in control of the entire food supply. Market forces, rather than national policies set by democratically elected officials, would control agricultural food systems. Under this scenario, each country would only produce a few export commodities, wiping out local food production, small family farms, and greatly compromising global food security. As a result, the human right to food would be dependent on multinational corporations the risk of hunger and famine worldwide. The stated goal of agribusiness interests in 'free trade' negotiations is to eliminate quotas and tariffs so that they can gain access to foreign markets for their products, called "market access." Since most agribusinesses are from rich countries in the global North, market access generally benefits those corporations at the expense of small farmers in poor countries.

The government instituted subsidies to farmers, actually benefit multinational corporations by keeping prices low while taxpayers foot the bill. Benefits of subsidies to corporate agribusinesses allow them to buy crops from small farmers at a price far below the cost of production and allow them to sell at market rates. Though, subsidies provide some relief to limited number of struggling family farmers, they are not enough to make up for low commodity prices but contribute to the erosion of small farms. This has led to a global depression in commodities prices. In fact, the prices that farmers receive today for their crops are likely to be well below the cost of production, causing millions of farmers to lose their land and others to go further and further into debt. The greatest threat to the livelihood of small farmers in both the developed and developing world is low commodity prices.

The policies driven by corporate globalization are pushing farmers off the land, and peasants out of agriculture. This is not a natural evolutionary process but a violent and imposed process. About more than 150,000 farmers' suicides is one aspect of this violence. The killing of dozens of peasants in Nandigram who were resisting land acquisition for a Special Economic Zone is another aspect of the violence involved in the forced uprooting of India's small farmers (India Needs Her Small Farmers, Vandana Shiva, 30th April, 2007). There are evidences that about seven thousand farmers have committed suicide in India in just three years. The Asian Age in its lead story on Sunday quoting government statistics said the average is even more frightening: it works out to six farmers killing themselves every day during the period. According to the Newspaper no mechanism has been put in place to monitor the progress of implementation of the Prime Minister's special package for distressed farmers in 31 suicide-prone districts of Andhra Pradesh, Karnataka, Kerala and Maharashtra.

Quoting Union Agriculture Ministry, the paper said the figure complete only till November 2006, reveals that 6,899 farmers have committed suicide till that date. Of the affected States Maharashtra, which was singled out by Prime Minister Manmohan Singh for a special visit, registered a tremendous increase in suicides by farmers with the toll rising from 577 in 2005 to 1,843, according to statistics till October 31, 2006. Kerala, according to government statistics, has registered an increase as well from 120 last year. Andhra Pradesh, which had witnessed 1,126 farmer suicides in 2004-05, brought the figure down to 92 last year, while Karnataka's highest of 78 deaths in 2003-04 has been reduced gradually to 45 till November last year. The paper quoted the Prime Minister relief fund to improving the plight of the farmers which include setting up dams, waiving off loans, *ex gratia* assistance from the Prime Minister's Natural Calamities Relief Fund of Rs. 50 lacks per district, an estimated credit flow of Rs. 1,275 crores for the financial year 2006-07 has very little effect on farmers' welfare.

Policy Implications

India is an agricultural country still does not have a comprehensive agricultural policy to enhance the food production and income of the farmers, especially small farm operators. Large number of farmers engaged in farming is not by choice but compulsion as they do not have other alternative source of livelihoods. There is a need to have accelerated effort to improve productivity, profitability and sustainability of small farmers to ensure food and nutritional security. There is a need to focus on farming system (livestock based farming system) rather than diversification of cropping system alone. Also, there is a need to provide adequate feed and fodder to over a billion farm animals as the country requires more than 200 million tonnes of milk to meet the nutritional security of growing population by 2030.

The critical issues in Indian agriculture are related to knowledge and infrastructure. In this emerging scenario of globalization and privatization of agriculture, government alone cannot handle the problems of small farmers. Hence, Public-Private partnership and institutional innovations like self-help groups, farmers clubs, etc., may be useful for harnessing the indigenous knowledge to improve performance of agriculture. Moreover, equal partnership can be developed with private sector to increase the access to market, technology, and global opportunities. Agricultural policies should be more farmer-centric than farm-centric only. In fact, farmers need to be empowered in decision making process. Moreover, adequate support in terms of investment in research, extension and training should be provided by the government.

Market distortions happen mainly because of the uncertainty with respect to government policies and, therefore, defining long term policy with respect to credit is very important. Since, the small farmers are vulnerable to the monsoons and unable to bear the risk they need focused attention for providing the services like credit and extending comprehensive crop insurance coverage to agricultural commodities which are perishable but of great social value due to their nutritional potentials. Investments in agriculture should promote local knowledge and organic and sustainable production systems rather than artificial fertilizers, pesticides, and herbicides.

A small farm model is the only way to achieve broad-based economic development, where poor people themselves are the source of production within an economy. In this context, land reform policy needs to be implemented with new initiatives. There can be no real sustainable development without implementation of effective land reform policy to help needs of millions of poor landless peasants. Priority must be given to fair and adequate redistribution of lands that have been concentrated from colonial times in the hands of an elite few. However, mere redistributing land is not enough but land reform policies must assure distribution of good quality land to truly needy families so that the small farmers' agricultural production becomes viable. Moreover, they should be given adequate support services like access to market, credits and good technical assistance about sustainable or organic kinds of production practices that provide them an opportunity to succeed. Tenancy laws should be modified to allow the large farmers to take up non-farm activities and to enable the marginal and small farmers to have greater access to land through lease market. Similarly, consolidation of holdings of even smaller holdings would enable them to enjoy better economies of scale.

Efforts should be made to create more employment opportunities outside agriculture to divert the surplus manpower from agriculture sector. Hence, promotion of agro-processing and service sector will be very important. However, stable power supply and necessary infrastructure will be necessary to attract private entrepreneurs. Farmers should be provided adequate protection in terms of procurement and minimum support prices to fruits and vegetable crops. Access to cold storage, market and transportation facilities to small farmers, particularly in backward regions is required. Besides, warehousing and storage facilities for foodgrains nearer to farmers' habitation, preferably at Block level need to be created.

Agriculture should be kept out of the WTO as food is a human right and should not be treated the same as any other commodity. Hence, governments may set national policies that encourage and guarantee food and nutritional security for all citizens. Government policies may establish supply management and food security reserves to prevent cheap commodities from being dumped on global markets. Market Access needs to be improved and developed countries should reduce high import tariffs on processed and semi-processed commodities. They may also put qualitative restrictions on imports as well as domestic subsidies for the protection of and support to household-subsistence farming. While developing countries may be encouraged to produce food for their domestic market.

Natural biodiversity needs to be protected through government intervention and seeds, plants, animals, and their components may be exempt from patenting. Agricultural policy must preserve the rights of Indigenous knowledge (as related to agriculture methods, use of seeds and plants) of farmers to utilize their cultural knowledge and collective use of resources may be protected from bio-piracy. The Trade Related

Aspects of Intellectual Property Rights (TRIPs) provisions in the WTO that permit multinational corporations to patent seeds originally developed by farmers, requiring farmers to pay for the right to replant those seeds, must be abolished.

Laws and regulations on sanitary and phytosanitary standards may guarantee high quality and safe food for consumers and the environment. Genetically Modified Organisms (GMOs) have yet to be proven safe. Utilizing the pre-cautionary principle, any trade agreement need to ban the trade of genetically-modified substances.

Conclusion

Food and nutritional security to a country like India is a major challenge. Despite substantial increase in food production due to green revolution, still more than 40 per cent of the population in India is malnourished. Land as the basic source of food production is highly skewed and varies to a large extent across various States. More than 85 per cent of the landholdings are small, owing less than 2 hectares of land, accounts for about 43 per cent of the cultivated area. The average size of holding is declining due to subdivision of landholdings. Development of small farmers is vital for the country as they can only provide food and nutritional security. Since farming is becoming less profitable large number of farmers are not willing to continue farming because of various constraints. The potential of Small holders can be realized only if they are empowered to have access to the crucial resources and entitlements, viz., land, water, energy, credit, insurance, and appropriate technologies. Policies and programmes need to focus on agriculture-led broad-based economic development by according highest priority to small-scale agriculture to enhance equity and sustainability of incomes and livelihoods. Fragmentation of farms below a certain size, say 0.5 ha, should be stopped. Land lease markets may be liberalized to promote scale economy. Government need to play a facilitating role to empower farmers and integrate agricultural policies with the rural development policies. *Panchayati Raj* Institutions may play an important role in raising awareness of the masses about these Rights. Multi-disciplinary agricultural research requires to become more development oriented with focus on the resource-poor small farmer. The role of the private sector in augmenting agricultural research, education and extension for development, particularly for welfare of the farmer must be encouraged.

REFERENCES

Bardhan, P. K. (1973): "Size, Productivity and Returns to Scale: An Analysis of Farm-level Data in India Agriculture", *Journal of Political Economy*, Vol. 81 (6): 1370-86.

Berry, R. A. (1972): "Farm Size Distribution, Income Distribution and the Efficiency of Agricultural Production: Colombia", *American Economic Review*, Vol. 62 (1): 403-08.

Berry, R. A. and W. R. Cline (1979): *Agrarian Structure and Productivity in Developing Countries* Baltimore: John Hopkins University Press.

Bhalla, S. (1979): "Farm Size and Productivity and Technical Change in India Agriculture" in R.A. Berry and W.R. Cline (ed.), *Agrarian Structure and Productivity in Developing Countries*, Baltimore: John Hopkins University Press, pp. 141-93.

Bhalla, S.S. and P. Roy (1988): "Mis-Specification in Farm Productivity Analysis: The Role of Land Quality", Oxford Economic Papers, Vol. 40 (1): 55-73.

Binswanger, Hans P. and Mark R. Rosenzweig (1986): "Behavioural and Material Determinants of Production Relations in Agriculture", *Journal of Development Studies*, Vol. 22(3): 503-39.

Chadha, G. K. (1978): "Farm Size and Productivity Revisited– Some Notes from Recent Experience of Punjab", *Economic & Political Weekly*, Review of Agriculture, Vol. XIII (39): A87-96.

Chand Ramesh; P.A. Laxmi Prasanna, and Aruna Singh (2011): Farm Size and Productivity: Understanding the Strengths of Small Holders and Improving Their Livelihoods, *Economic and Political Weekly Supplement*, June 25, 2011, Vol. XLVI, No. 26 and 27.

Chand, Ramesh (2010): "Achieving 4 per cent Growth Rate in Agriculture during the Eleventh Five-Year Plan" in Pulin B. Nayak, Bishwanath Golder and Pradeep.

Chengal Reddy (2009): The Role of Farmers Organsiations in Pro-Poor Policy Process, CIFA, presentation at Belgian Technical Cooperation (BTC) Seminar on Agriculture as a Motor of Pro Poor Growth, Brussels, 16-17 December 2009.

Chattopadhyay, Manabendu and A. Sengupta (1997): "Farm Size and Productivity – A New Look at the Old Debate", *Economic & Political Weekly*, 27 December, Vol XXXII (52): A 172-75.

Deolalikar (1981): "The Inverse Relationship between Productivity and Farm Size: A Test Using Regional Data from India", *American Journal of Agricultural Economics*, Vol. 63 (2): 275-79.

Dyer, Graham (1997): *Class, State and Agricultural Productivity in Egypt:*

A Study of the Inverse Relationship between Farm Size and Land Productivity, London: Frank Cass Publishers.

Fan, Shenggen and Connie Chang-Kang (2005): "Is Small Beautiful? Farm Size, Productivity and Poverty in Asian Agriculture", *Agricultural Economics*, Vol. 32 (Supplements-1): 135-46.

Foster, Andew D. and Mark R. Rosenzweig (2010): "Is There Surplus Labour in Rural India?", Centre Discussion Paper No. 991, Economic Growth Centre, Yale University.

Frisvold, George B. (1994): "Does Supervision Matter? Some Hypothesis Tests Using Indian Farm Level Data", *Journal of Development Economics*, Vol 43 (2): 217-38.

Ghose, A. K. (1979): "Farm Size and Land Productivity in Indian Agriculture – A Reappraisal", *Journal of Development Studies*, Vol. 16 (1): 27-49.

GoI (2003) and (2005): NSSO 59th and 60th Round, Ministry of Statistics and Programme Implementation, Government of India, New Delhi.

Haque, T. (1992): Economics of Agriculture in Backward Regions, *Project Report*, NIRD, Hyderabad.

Haque, T. (1985): Regional Trends and Patterns of Diversification of the Rural Economy in India, *Indian Journal of Agricultural Economics*, Vol. XL, No. 3, July-Sept., pp. 291-97.

Hanumantha, Rao C. H. (1975): *Technological Change and Distribution of Gains in Indian Agriculture*, New Delhi: Macmillan.

——— (1966): "Alternative Explanations of the Inverse Relationship between Farm Size and Output Per Acre in India", *Indian Economic Review*, Vol. 1(2).

Havenevik, J. Kjell and Rune Skarstein (1997): "Land Tenure, State-Peasant Relations and Productivity in Tanzanian Agriculture" in Amit Bhaduri and Rune Skarstein (ed.), *Economic Development and Agricultural Productivity*, Cheltenham: Edward Elgar, pp. 183-212.

Hazell, Peter (2011): "Five Big Questions about Five Hundred Million Small Farms", keynote paper presented at the IFAD conference on New Directions for Smallholder Agriculture, 24-25 January 2011, International Fund for Agricultural Development, *via* Paolo Di Dono, 44, Rome 00142, Italy.

Khusro, A. M. (1968): "Returns to Scale in Indian Agriculture" in A.M. Khusro (ed.), *Readings in Agricultural Development*, New Delhi: Allied Publishers.

Mazumdar, Dipak (1965): "Size of Farm and Productivity: A Problem of Indian Peasant Agriculture", *Economica*, Vol. 32, May, pp. 161-73.

Oksana, Nagayets (2005): "Small Farms: Current Status and Key Trends", information brief prepared for workshop on the Future of Small Farms Research, Wye College, London.

Raghbendra J., P. Chitkara and S. Gupta (2000): "Productivity, Technical and Allocative Efficiency and Farm Size in Wheat Farming in India: A DEA Approach", *Applied Economic Letters*, Vol. 7(1): 1-5.

Rudra, Ashok (1968): "Farm Size and Yield per Acre", *Economic & Political Weekly*, Vol. III (43): 1041-44.

Sen, A. K. (1962): "An Aspect of Indian Agriculture", *The Economic Weekly*, Vol. 14 (4-6): 243-46.

——— (1964): "Size of Holdings and Productivity", *The Economic Weekly*, Vol. 16 (5-7): 323-26.

Saini, G. R. (1971): "Holding Size, Productivity and Some Related Aspects of Indian Agriculture", *Economic & Political Weekly*, Review of Agriculture, Vol. VI (26): A 79-A85.

Singh, R. B. (2011): Towards An Evergreen Revolution: The Road Map, National Academy of Agricultural Sciences, New Delhi.

Swmainathan, M.S. (2012): 10th Convocation Address at NDRI, Karnal, February 21, 2012

Subbarao, K. (1982): "Technology Gap and the Emerging Size-Productivity Relationships Following the Adoption of New Technology: An Analysis of Evidence from North-West and Eastern India", unpublished paper, University of California at Berkeley, Department of Agriculture and Resource Economics.

Thapa, Ganesh and Raghav Gaiha (2011): "Smallholder Farming in Asia and the Pacific: Challenges and Opportunities", paper presented at the IFAD conference on New Directions for Smallholder Agriculture, 24-25 January, 2011, International Fund for Agricultural Development, *via* Paolo Di Dono, 44, Rome 00142, Italy.

Vandana, Siva (2007): India needs her small farmers, 30 April, 2007.

ANNEXURE 6.1

Average Size of Operational Holdings (Hectares) by Major Size Groups, 2000-01

(Hectares)

State/UT	*Marginal*	*Small*	*Semi-Medium*	*Medium*	*Large*	*All Holdings*
Andhra Pradesh	0.44	1.42	2.67	5.70	16.34	1.25
Bihar	0.30	1.21	2.62	5.24	15.50	0.58
Chhattisgarh	0.44	1.42	2.70	5.76	16.49	1.60
Goa	0.32	1.26	2.56	5.64	23.77	0.84
Gujarat	0.53	1.46	2.78	5.80	16.91	2.33
Haryana	0.45	1.43	2.81	5.99	16.48	2.32
Himachal Pradesh	0.41	1.40	2.71	5.69	15.91	1.07
Jammu & Kashmir	0.37	1.40	2.66	5.39	21.13	0.67
Karnataka	0.46	1.44	2.72	5.83	14.83	1.74
Kerala	0.14	1.32	2.52	5.29	40.93	0.24
Madhya Pradesh	0.49	1.45	2.77	5.94	15.50	2.22
Maharashtra	0.50	1.42	2.69	5.64	15.38	1.66
Orissa	0.50	1.39	2.69	5.63	16.48	1.25
Punjab	0.63	1.40	2.67	5.75	15.14	4.03
Rajasthan	0.48	1.44	2.85	6.19	18.21	3.65
Tamil Nadu	0.37	1.40	2.72	5.68	19.48	0.89
Uttrakhand	0.39	1.39	2.71	5.47	25.07	0.95
Uttar Pradesh	0.40	1.41	2.74	5.57	15.07	0.83
West Bengal	0.51	1.59	2.77	5.12	**278.95**	0.82
Pondicherry	0.29	1.42	2.74	5.68	19.50	0.70
All-India*	**0.24**	**1.42**	**2.39**	**4.42**	**13.16**	**1.33**

Note: Includes institutional holdings also.

* No data available for Jharkhand.

Source : Department of Agriculture and Cooperation, Agricultural Census Division.

ANNEXURE 6.2

Incidence of Indebtedness according to Farm Size in various States of India

Sl. No.	*State*	*% of indebted Farmer Households*				
		Marginal farmers	*Small farmers*	*Semi-medium farmers*	*Medium farmers*	*Large farmers*
1.	Uttar Pradesh	71.3	17.4	7.8	3.4	0.3
2.	Maharashtra	36.0	26.2	23.3	12.2	2.4
3.	Madhya Pradesh	33.0	27.1	23.1	13.0	3.9
4.	Rajasthan	43.9	19.8	17.8	14.1	4.5
5.	Karnataka	50.7	22.8	15.9	9.3	1.2
6.	Andhra Pradesh	55.7	21.8	15.1	6.6	0.7
7.	Bihar	86.9	9.2	2.8	0.7	0.6
8.	West Bengal	88.7	8.5	2.4	0.4	0.0
9.	Punjab	53.3	15.8	17.0	11.8	2.2
10.	Orissa	70.3	20.6	7.3	1.7	0.0
	All India	61.0	18.9	12.5	6.4	1.2

Source: Report No. 498(59/33/1), Situation Assessment Survey of Farmers: Indebtedness of Farmer Households, National Sample Survey 59th Round (January-December 2003)

ANNEXURE 6.3

State-wise Number of Kisan Credit Cards issued to Farmers

(Upto 30th November 2009)

Sl. No.	*State/UTs*	*Percentage share of various agencies in credit cards issued (%)*				*Total Number of credit cards issued (in thousand)*
		Co-operative Banks	*Regional Rural Banks*	*Commer-cial Banks*	*Total*	
			Percentages			
1.	Uttar Pradesh	38.78	22.35	38.86	100	15,894
2.	Andhra Pradesh	24.50	13.58	61.92	100	14,673
3.	Maharashtra	65.75	4.02	30.23	100	8,201
4.	Tamil Nadu	30.35	4.89	64.76	100	5,939
5.	Madhya Pradesh	60.10	10.08	29.83	100	5,374
6.	Karnataka	33.92	22.61	43.47	100	5,323
7.	Orissa	66.49	12.95	20.57	100	5,298
8.	Rajasthan	59.29	9.72	31.00	100	4,899
9.	Kerala	43.15	14.00	42.86	100	3,321
10.	Bihar	24.03	29.38	46.58	100	3,312
11.	West Bangal	46.76	10.77	42.47	100	3,204
12.	Gujarat	41.68	8.63	49.69	100	2,879
13.	Haryana	52.29	14.88	32.82	100	2,402
14.	Punjab	39.53	5.74	54.72	100	2,276
15.	Chhattisgarh	64.20	18.53	17.27	100	1,580
16.	Jharkhand	26.56	33.41	40.03	100	1,050
17.	Uttaranchal	48.53	7.68	43.78	100	618
18.	Assam	2.96	27.98	69.06	100	485
19.	Himachal Pradesh	34.63	11.62	53.75	100	420
	Total	**48.20**	**12.27**	**39.52**	**100**	**87,151 (100)**

Source: Department of Agriculture and Cooperation, Credit Division.

7

Agrarian Change in Mountain Economies: The Constraints and Policy Options

I.C. Awasthi

Introduction

The mountain region covers 10 States namely Jammu & Kashmir, Himachal Pradesh, Uttarakhand, Sikkim, Arunachal Pradesh, Nagaland, Manipur, Mizoram, Tripura, Meghalaya and Darjeeling district of West Bengal and Karbi Anglong and North Cachar districts of Assam.

The mountain economies has increasingly confronted with the challenges and problems concerning to sustainable livelihoods, mountain agriculture, persistent poverty, fragile environment, and female drudgery. Traditionally, hill people have been meeting their food and other non-food requirements through crop cultivation, livestock rearing and collection of forest produce. Increasing population along with rising aspirations of people due to expansion of development activities and better livelihood options has led to increasing pressure on land and other limited traditional resources. The cultivable land being scarce getting further fragmented with the increase of population and it has hard pressed the people to look for alternative livelihood strategies.

The alternative strategies consisted of extension of cultivation on hitherto uncultivated marginal land (though limited), improvement in quality of livestock, diversification of agriculture in to high value crops such as horticulture and non-food crops for the market. But these coping strategies have limitations to provide food security to the people on a sustainable basis due to low productivity of resources and the problems associated with the hill specificities on the one hand and increasing population on the other. The limited resources, low bargaining strength, lack of appropriate institutional mechanism for access to market, credit and other inputs has severely restricted expansion of livelihood opportunities. The mountain agriculture, therefore, could not ensure food security and instead increased the vulnerability of the people.

Also, over the years pressure on land has resulted degradation of pastures and erosion of common grazing land. Horticulture (fruits and vegetables) has potential for high returns but technological and other constraints have severely restricted the scope of expansion. The agriculture is largely traditional with little diversification to potential areas according to its land capability. Improvements in agriculture technology elsewhere did not help much to the hill economy due to problems associated with hill specificities.

Though, there has been some diversification from cereal growing to horticulture (mainly fruits and vegetables) over a period of time yet it was limited to certain pockets of the economies that are generally well connected to road transport. Overall, there has been insufficient food supply within the mountain region and dependence on food imports from outside region increased. The preconditions of agriculture growth, which critically depends on the intensification of resource use, input absorption capacity, infrastructural back up and economies of scale, do not exist in the hill region due to the conditions of fragility and marginality. Also, relatively low accessibility results in limited mobility and high cost of inputs, which severely restricts the possibility of promoting staple or chief commodity and exploiting other niche products in the region. This in turn inhibits trade and specialisation to grow for market principally because neither surplus is produced nor the hill specificities permit trade to develop in a wider scale. In any case trade in the region is minimal for the market primarily because of subsistence nature of economy.

The manufacturing base in any case is miniscule in these areas; no large industries can be set up in the hilly region due to fragility;, lack of transportation network (rail), markets, power and economies of scale. Village and household industries based on livestock, food processing and natural resource have been promoted but it could not generate self-reinforcing and self-sustaining forces due to small scale operation, little quantity, dispersed production and lack of access to appropriate institutions. Over a period of time the base of these industries have weakened and deteriorated. Moreover, these industries have been subjected to increasing competition from the similar products from the plain areas that are superior in terms of quality and price much cheaper than the local products. This has had an adverse impact on the local economy.

When agriculture and industrial base is weak this would obviously have far reaching implications to the hill economies. The development strategies, therefore, requires to be distinct in different hill economies in view of regional specificities that need to be focused and promoted on mountain resources that has clear comparative advantages (hydro power, tourism, horticulture, and other niche products).

This chapter has been structured into four sections. The first section deals with the characteristic features of mountain economies and second section describes the diversity of development patterns and growth trajectory. The third section briefly overviews the agrarian and livelihood situation in one of the mountain States namely, Uttarakhand based on empirical evidences and the final section summarizes the main arguments and reports policy imperatives.

Diversity in Development Patterns

In spite of similar biophysical attributes among all the hill regions, the development patterns vary across the hill States in terms of various development indicators.

The case of Himachal Pradesh provides unique example among all the hill States, the transformation from subsistence agriculture to one of the lead horticulture-producing region is worth emulating. The development strategy gave primacy to development of transport network (mainly, roads), communication and hydroelectric projects. Government provided institutional support for diversifying from the traditional cereal crops to horticulture crops along with dairy development. The State also adopted an appropriate land-use pattern with a view to optimizing the returns. This has significantly improved livelihood options and the income levels of the people. Health and education sector have shown impressive growth. The forests have been exploited to produce mainly timber and resin that contributes to about 35 per cent from the State (Negi, *et al.*, 2004).

In industry, emphasis is being laid on setting up of industrial units, which are environment-friendly and based on local raw material and niche products. Industrial sector contributes nearly 15 per cent of GDP. Some of the leading industrial houses have been attracted to set up industrial units in the State and introduction of single window clearance for medium and large-scale sector has facilitated promotion of industrial units. Sustainable tourism industry has been promoted without disturbing ecology and environment. Various types of tourism is being promoted in the State such as pilgrimage, tribal, health, adventure sports, wildlife and cultural primarily to provide employment and incomes. State is endowed with vast potential of hydropower. The estimated potential is about 20,392 MW of which 6,045 MW has been so far harnessed. The State is one of the most socially progressive States in the country. It has had lowest decadal population growth after Nagaland and Sikkim, highest literacy rates next to Mizoram and Tripura and low poverty levels. It has performed better in terms of human development indicators and has substantially higher per capita consumption expenditure (second highest in the urban areas) with lower inequality of per capita expenditure. The State has highest per capita Net State Domestic Product (NSDP) and one of the fastest growing hill economies. Employment growth has substantively increased but unemployment rate continue to increase unabated.

Despite the similar bio-physical characteristics, the development patterns vary across the hill States in terms of various development indicators. Himachal Pradesh provides one of the notable examples in the transformation process from traditional cereal crop producing to one of the lead horticulture producing regions. Himachal Pradesh has made considerable progress in terms of high literacy rates, almost universal enrolments at primary level, availability of schooling facilities and expansion in health care and physical infrastructure. In several aspects, the State has established benchmarks that have become role models for the other hill States of the country. Focused public interventions, greater participation of women and strong local institutions have been supportive policy measures. Notwithstanding, the State has issues of concerns that need to be addressed. It has huge potential for hydro-power, tourism, excellent agro-climatic diversity for value adding agro business. The State has formulated comprehensive framework for creating investor friendly environment to ensure sustainable industrial development.

Performance of Jammu and Kashmir, on the other hand, does not seem to be that encouraging as in the case of Himachal Pradesh. The State has recorded relatively highest decadal population growth after Meghalaya and lowest overall and female literacy rates and highest gender disparity among all the hill States. Though the poverty level is the lowest after Nagaland yet per capita NSDP is low, lower than national average. The infant mortality rate (IMR) is estimated at 108 per thousand. In spite of a large natural and human resource base, the State lags behind in industrial development. The State possesses immense potential for development of the power sector but the performance has not been on a scale commensurate with the possibilities, only a mere 10 per cent of the total potential has so far been harnessed. In terms of road length per 1000 sq. km, the State ranks the lowest (127/1000 sq km) as compared to Himachal Pradesh (542/1000sq km) and Assam (872/1000 sq km). Also, the inter-district variations in accessibility of roads are very sharp. For example, in terms of road length per 100 sq. km., Budgam district has the highest road density of 81.84 km and the districts of Leh, Kargil and Doda have poor accessibility to the tune of 2.58, 4.82 and 5.24 km respectively (Sadhu, *et al.*, 2004). Tourism has been the major economic activity of the State but due to insurgency within the valley since 1989 tourist inflow received a major setback. State has not made any concerted efforts for tourism diversification. The area under fruit and vegetable cultivation has increased marginally though there is immense potential for growing horticulture crops, vegetables, floriculture and medicinal herbs. There is also good potential to promoting handicrafts and handlooms industry.

Jammu & Kashmir has the one of the lowest poverty ratio among all the hill States and relatively better natural resource base but lags in industrial development. The social indicators, in particular gender discrimination in education, infant mortality is still very high. State has immense potentials for tourism and hydro power. It offers vast scope for horticulture production and has unique climate for cultivation of saffron and suitable conditions for floriculture, aromatic and medicinal plants. But these resources have not been harnessed to the extent desired. There is also good potential for promoting handicrafts and handlooms industry. However, strained Indo-Pakistan relations and continued insurgency in the State has created major setback.

Uttarakhand being new State (carved out from the Uttar Pradesh in 2000) and the database for the independent State is not that robust as in the case of other hill States. However, some of the aggregate indicators that are available point to the state of economy. Literacy rates are impressive with relatively lower gender gap. Poverty ratio has sharply declined in recent year but urban poverty is still higher than rural poverty. State has performed relatively better in terms of development. One of the principal reasons for lower performance is that the State failed to incorporate mountain specific development as was in the case of Himachal Pradesh, such development specificity has been by and large found missing in Uttarakhand (Pokhriyal, 2004). Though there are good potentials for certain products, in which Uttarakhand is supposedly to have advantage, like hydro-electricity, wool processing, forest produce, horticulture and tourism and amenities (Awasthi, *et al.*, 2005). However, many of these activities could not be developed primarily due to lack of appropriate development strategy. For instance, the estimated potential of power generation is about 20,450 MW in the State of which 11.0 per cent could be harnessed till 2007. There is a huge gap between industrial activities in the plains and hills primarily because of critical constrains in the latter case. Recent Industrial Policy (2008) in particular addresses the economic development in the hills where industrial activities has been by and large considered to be dormant.

People, particularly at a higher altitude, have traditional skills and technology and have been producing wool and woollen products for market either in exchange of commodities or for money since long time but now by and large this home based activity is obliterating. Likewise, the region is well known for Hindu pilgrimage, famous glaciers, trekking, and expedition places and offer potential for development of tourism as an industry. However, the linkages of tourism industry within local economy are found to be extremely weak.

Uttarakhand has recorded impressive literacy rates with relatively lower gender gap. Though there are good potentials for hydro-electricity, tourism and amenities, horticulture, wool processing, forest produce. However, these resources have not been harnessed due to lack of economic infrastructure and institutional support. Recent economic indicators, however show much better performance in terms of sectoral growth and other development indicators in the State.

North-East States present somewhat mixed pattern in development front. "The trauma of partition, political evolution and insurgency combined with the geographic allocation, transport bottlenecks, natural calamities, etc., have hindered the progress of the North-East, and the region has experienced relatively slower economic growth compared to the rest of the country (Planning Commission, 2002)".

Most of the States have still very high rate of population growth, though declined considerably during last 2 decades, yet most of the States have been growing much higher than national average except in the case of Tripura. Literacy has improved significantly over the years, yet it is still lower in Arunachal Pradesh compared to all India, however the gender gap in literacy is lower in comparison to other hill States. Poverty has declined in recent years but Manipur has highest ratio of poverty (47 per cent). NSDP growth in most of the States has been lower than the all India average. Unemployment rate is huge and is higher than national average. Income inequality, measured by gini-coefficient, has widened in most of these States in recent years barring Arunachal Pradesh and Manipur.

Primary sector still constitutes the bulk of the labour force in all the hill States of the region and the secondary sector has remained stagnant. In some cases, the hill economies are experiencing some sort of deindustrialisation (Das, *et al.*, 2004). The decline in the share of the primary sector is being compensated by the rise in the share of tertiary sector and the rise in the share of public expenditure has been the cause behind the growth of the share of the tertiary sector. The economic development in these economies is yet to be rooted in self-sustaining process (Das, *et al.*, 2004). Sikkim, despite its impressive performance in economic and social indicators, the poverty levels and unemployment rates continues to be significantly higher and sustainability of major livelihood options (agriculture crops) in which most of the poor depended are declining or stagnating in terms of productivity levels (Papola, 2003).

Efforts to promote horticulture and cash crops as a strategic intervention to replace shifting cultivation by the government has however been mixed and have failed to completely replace shifting cultivation. Agricultural transformation through horticulture and cash crops has raised social, economic and environmental related problems. Such transformations to horticulture and cash crops failed to satisfactorily deliver the promised economic returns, conversion to settled agriculture has increased vulnerability and has eroded the self-reliance hitherto inherent in upland agriculture (Choudhury, 2004). One of the reasons is that land management is the most neglected part of the development policy and planning of this region (Maithani, 2004). Increasing proportion of rural households with no cultivable land in the region (except in Manipur and Meghalaya) and disproportionately rising small and marginal farmers have added constraints in livelihood in rural areas (Das, 2004).

Being 'special category' States, they receive 90 per cent grant and 10 per cent loan and of these assured central assistance, they do not make any efforts to mobilize their own resources. This has resulted in increasing dependence on central resources. Various indicators show that all the hill States of North-East come under 'very low' category in overall development except Manipur, which is under 'low' category (Goswami and Gogoi, 2004).

In all the North-Eastern States including other hill States priority to industrialization is low; it forms a small part of total outlay. Health allocation is also low, in most of the States. North-East States have remained industrially most backward region. Decadal population is still very high in majority of States (Nagaland, Mizoram, Meghalaya, Arunachal Pradesh and Manipur) compared to all-India average. Per capita income is lower than the national average. North-East region continues to be the net importer of foodgrains. Bulk of labour force is engaged in agriculture and allied activities with high unemployment rates, particularly among youth.

Most of the hill States have good potential for development particularly in horticulture, hydro power generation, tourism and other location specific products. The region has also good potential for petroleum and natural gas resources (North-Eastern region). Despite of good resource base, there is acute deficiency of critical economic infrastructure in the region. There is need for creating institutions and building capacity in order to unleash and spur development activities.

Subsistence Agriculture and Livelihood Diversities: Case of Uttarakhand

Uttarakhand hill region is characterised by subsistence agriculture and any drastic transformation within cereal-based agriculture on terraced type cultivation has obvious limitations to expand its land-based activities. The industrial base, in any case, is precariously small and large-scale industrial activity is neither possible nor desirable from the point of view of ecological and environmental considerations. From long term perspective, traditional agriculture with its low productive base and rudimentary industrial activities does not seem to be a viable and/or sustainable option for providing employment and livelihood to the burgeoning labour force. This precisely makes a strong case for diversification within agriculture sector according to its potential land capability and also for promoting other non-farm pursuits.

(i) Multiplicity of Employment

People deriving income from different sources are the most common features in poor regions as single source of income is hardly sufficient to eke out living. It is common to our understanding that poor regions are generally characterized by underdevelopment in terms of its resource base both physical and human and various factors reinforces it to remain at low equilibrium trap in development trajectory. This is more so in agrarian economies, which have abundance of labour supply that is manifested in terms of huge unemployment and underemployment in various forms. Lewisian development framework precisely deals with this kind of phenomenon and it is argued that industrialization process would eventually help to graduate backward economy into development process with near full employment.

However, such development hypothesis has not proved correct and people in poor regions continue to have their dependence on multiple sources of income for living, despite the fact that there has been industrial development in other regions. Engagement in multiple sources of livelihoods is most common not only among poor but relatively well-off households in order to augment their incomes. The two different processes seem to work simultaneously; first as a deliberate household strategy to enhance income as an involuntary response to crisis situation (Stark, 1991) and secondly to seize opportunities as a means of accumulation (Davies, 1996). In poor regions people depend on various sources of employment primarily because of sustenance in the absence of any sole option of employment. Multiple economic activities are most pervasive and widespread among different groups of population in the survey areas in Uttarakhand, manifesting symptoms of backward economy.

It can be observed that about 26 per cent among all the workers have single or sole activity while a large majority of them (74%) resort to more than one activity (Table 7.1). It is most common feature in all the Blocks and districts in the hill region of the State. Apparently, there does not seem to be operating a single pattern of activities in different workers' group, rather different processes seem to work simultaneously across that govern the plural-activity of work pattern. These processes are governed by the opportunities and options of livelihood available to the population by operating in several labour markets. Interestingly, landless workers seem to be depending primarily on the single activity (68%), probably because the lack of opportunities and alternative options available to them due to entry barriers in the labour markets. This could precisely be the reason for lowest household incomes in this group.

While in the case of marginal and small landholding classes, dependence on multiple activities increases in order to augment their household income. However, in the medium landholding class such dependence on multiple activities declines, probably because household have relatively higher incomes from agriculture and animal husbandry and other assured sources of incomes such as from pension and other

Table 7.1: Multiplicity of Employment among All Workers

Household feature	*One employment*	*Two employment*	*Three employment*
Social Category:			
Land Class (acre)			
Landless	67.44	23.26	9.30
Marginal	24.68	51.56	23.76
Small	19.63	52.34	28.04
Medium	27.78	50.00	22.22
Income group (MPCI)			
Up to Rs. 250	19.35	62.37	18.28
Rs. 250 to 500	23.58	55.40	21.02
Rs. 500 to 1000	29.82	46.86	23.32
Rs. 1000 to 2500	24.70	47.32	27.98
More than Rs. 2500	22.58	51.61	25.81
Educational levels			
Illiterates	25.44	52.96	21.60
Primary	19.08	50.00	30.92
Middle	24.48	50.21	25.31
High school	29.73	49.81	20.46
Intermediate	31.01	47.47	21.52
Graduate and above	22.00	54.00	24.00
Technical	20.00	30.00	50.00
Development status of Blocks			
Less developed	27.70	52.62	19.67
Developed	23.92	48.77	27.31
Total	25.76	50.64	23.61
Number of workers	324	637	297

Source: Field Survey, 2004-05.

incomes. Monthly per capita income (MPCI) group shows inverted U-shape curve, initially dependence on singular employment rises with rise in income levels but later tapers off as income levels rises. In the higher income group dependence on single employment declines and that of multiple employments rise. Educational levels also indicate somewhat similar pattern, initially dependence on single employment increases as we move to higher levels of education (up to intermediate) but then such dependence gradually declines and multiplicity of employment rises.

Such a huge dependence on multiple activities seems to be common feature in the mountain economy where cultivation is not only seasonal but the income from this activity is low (20%) that is not enough to provide livelihood beyond a few months in a year. Animal husbandry, though more or less perennial activity, at best supplements the household incomes. Therefore, multiple engagements in activities form an important part of household strategy. Though there does not seem to be perceptible variations in the activities pursued in the developed and less developed Blocks yet developed Blocks showing greater intensity of multiplicity of activities. This kind of phenomenon indicates relatively better livelihood opportunities available in the developed Blocks.

This finding of present study dispel the commonly derived hypothesis that poor households derive their income from multiple sources as part of their survival strategy while the relatively better off households and individuals depend on few and stable sources of incomes (Ellis, 2000; Gordon and Craig, 2001). It can be argued that in relatively less developed regions, lack of opportunities constrict the people to enter in the labour market while in the relatively developed region people seize the existing opportunities for better remunerated activities.

Extent of multiplicity of employment increases sharply among principal workers. About 92 per cent principal workers are such who are engaged in more than one type of employment. In the rural areas such multiplicity of activities is not unusual phenomenon and almost every principal worker is engaged either in cultivation or in animal husbandry activity.

(ii) Average Days of Employment

Average annual days of employment available in agriculture and allied activities are indeed very low in the hill areas. This is primarily because the nature of hill agriculture that poses insurmountable problems in terms of small, fragmented and scattered land holdings where only limited traditional crops can be had in a year and there is little scope for expansion either in terms of area or bringing about radical technological transformation. Irrigation facilities are very low and agriculture is mainly rainfed and use of HYVs and other chemical fertilizers is very low resulting in low productivity and low returns from agricultural crops. All these factors lead to limited food availability hence the food insecurity that is common feature in many parts of the hill districts in the State. Although, people seem to be working very hard in agriculture activities yet it does not provide employment beyond few months in a year. On an average agriculture and allied activities provide employment to about 56 days in a year for males while for females it provides for about 106 days employment (Table 7.2). Generally, in the context of hill agriculture animal husbandry provides more days of employment than agriculture particularly for females. This is principally due to the fact that agriculture is seasonal activity while the animal husbandry is perennial activity throughout the year. It is quite intriguing to observe that despite heavy engagement in agriculture and allied activities it provides employment not beyond two months for males and three and half months for females. Across development Blocks one does not find any perceptible variations. However, it can be observed that the relatively developed Blocks have generally higher person day's employment compared to less developed Blocks.

Table 7.2: Average Annual Person Days of Employment Per Worker

District	*Block*	*Agriculture and allied (days)*	
		Male	*Female*
Almora	Hawalbagh	57	118
	Salt	53	108
Pithoragarh	Dharchula	56	114
	Berinag	55	102
T. Garhwal	Chamba	63	112
	Kirtinagar	51	99
Uttarkashi	Bhatwari	60	88
	Dunda	54	98
Social category	Brahmin	56	108
	Kshatriya	59	113
	SC	45	80
	ST	58	87
	Others	55	94
Land class	Landless	35	70
	Marginal	55	105
	Small	63	114
	Medium	74	118
Age-group	0-5	—	—
	6-14	25	40
	15-29	41	100
	30-59	70	121
	59+	62	93
Development	Developed	59	109
Status of Blocks	Less developed	54	102
Total		56	106

Source: Field Survey, 2004-05.

The reasons for higher days of employment in the relatively developed Blocks could be due to more areas under horticulture crops leading to commercialization of agriculture (vegetables, fruits and other non-food crops) compared to relatively less developed Blocks (Table 7.3). This supports our hypothesis that diversification into high value crops promotes more employment and incomes. Across caste category, scheduled caste has the lowest person day's employment available in agriculture and allied activities both for males and females and highest being for upper castes Kshatriya and Brahmin. Scheduled Castes (SCs) have the lowest days of employment available in agriculture and allied activities primarily because of lowest asset base (land, for example) that makes difficult to enhance their endowment base hence provide limited days employment in agriculture and animal husbandry. STs have relatively higher day's employment available, particularly for males, despite the fact that they have also lower per capita land available for cultivation. The reason for this is that STs are generally settled in higher altitude where agro-pastoralism is dominant and certain horticultural crops such as potato, radish, apple, *razma* (kidney beans) are grown in abundance along with animal husbandry (herd of sheep and goats) that provides raw wool for making woollen garments, which is the traditional occupation of the tribal community. Availability of days of employment and size of land class are positively correlated, higher the land size class the higher is the availability of man days and rightly so. This clearly indicates that land endowment is the most important factor in determining the intensity of employment of a worker. Higher size of land obviously helps to employ more number of people, whatsoever the level of productivity of people engaged in it.

Across age groups, days of employment increases gradually in the higher age group, lowest being in the age-group 6-14 and highest in the age-group 30-59 but then drops in the highest age-group 59 and above and rightly so. It is common in the hill districts that even school going children (age-group 6-14) help their parents in agriculture and allied activities, particularly in harvesting season when the pressure of work increases. Female children are more conspicuously engaged than their male children counterparts. While agriculture being the seasonal activity, the animal husbandry is perennial one that requires constant care and rearing animals. Children perform various activities such as taking animals for grazing, giving fodder and water, cutting fodder and bringing from the fields, cleaning the place of stay of these animals and sometimes milching the cows and buffaloes, etc. It is also observed that relatively developed Blocks have higher intensity of employment than the less developed ones, albeit marginally. The intensity of employment, therefore, (per person annual person days of employment) is very low in the hill areas of Uttarakhand.

Table 7.3: Area under Horticultural Crops (Acres)

District	*Block*	*Gross cropped area*	*Area under horticulture*	*Per cent*
Almora	Hawalbagh	64.33	23.89	27.08
	Salt	85.85	0.00	0.00
Pithoragarh	Berinag	72.57	5.47	7.00
	Dharchula	61.10	12.50	16.98
T. Garhwal	Chamba	20.75	18.41	47.01
	Kirtinagar	46.63	2.66	5.39
Uttarkashi	Bhatwari	12.00	24.83	67.42
	Dunda	32.62	1.95	5.64
Total		395.85	89.71	22.66

Source: Field Survey, 2004-05.

Average days of employment in casual works turns out to be about 51 days for males and 43 days for females per annum (Table 7.4). It is very low indeed and the nature of employment is such that people have to find multiple sources of employment to cope their living. Casual employment is mostly available in construction activities both in private and public domain. Most common activities are construction of individual houses and construction of roads, culverts, bridges and buildings in public works. However, it needs to be mentioned here that such jobs in the construction sector are not necessarily available within the boundaries of villages. The fact is that some of these jobs are available through daily commuting to nearby towns and rural bazaars. However, the access to such jobs for females is somewhat limited owing to various socio-economic barriers and inhibitions. It is interesting to note that days of employment in casual works is higher in relatively in less developed Blocks and female casual employment in particular is distinctly visible only in less developed Blocks. For females, casual employment in construction activities is not very encouraged due to social and cultural inhibitions, however, casual works in agricultural activities is most common for landless and poor families. However, agriculture and allied activities does not ensure such employment beyond few days in weeding and harvesting season and casual employment in construction then becomes economic necessity particularly for poor and landless families. It can be seen that females are employed in construction activities only in two Blocks, which are relatively less developed varying from 30 to 45 days employment.

Across social category, the dependence on casual employment is highest (in terms of days of employment available) among SCs (both males and females) followed by Kshatriya and Brahmin in that order and STs, on the other hand, have lowest days of employment available. One of the reasons for higher and lower dependence on casual type of employment is lack of assets (land, for example), low educational levels and non-availability of other regular sources of earnings that forces people to go for casual type of jobs involving hard physical labour. Also, availability of such employment is important, and such opportunities are most likely available in rural bazaars and nearer to Block and district headquarters. Proximity to such centres is perhaps the most important factors that generate variety of development activities and earning sources in construction and related activities. Dependence on casual works is higher among the sub-marginal, landless and marginal landholding classes and dependence on such employment gradually declines in the higher land size classes and the fact that none from the medium land holding class is

Table7.4: Average Annual Person Days of Employment Per Worker

District	*Block*	*Casual employment (days)*	
		Male	*Female*
Almora	Hawalbagh	59	–
	Salt	42	30
Pithoragarh	Berinag	62	–
	Dharchula	43	–
T. Garhwal	Chamba	53	–
	Kirtinagar	54	45
Uttarkashi	Bhatwari	40	–
	Dunda	57	42
Social category	Brahmin	42	25
	Kshatriya	53	40
	SC	56	49
	ST	33	–
	Others	37	–
Land class	no land	48	35
	>0 <= .50	52	44
	>.50 <= 2.50	52	–
	>2.50 <= 5.0	45	–
Age-group	0-5	–	–
	6-14	–	–
	15-29	47	44
	30-59	51	43
	59+	56	34
Development status of Blocks	Developed	49	–
	Less developed	53	43
Total		51	43

Source: Field Survey, 2004-05.

available for casual type of employment. No one is reported having engaged in casual employment up to age 14 years and the days of casual employment for males increases with higher age-groups while declines for females.

(iii) Income and Earnings

Income is the outcome of diverse portfolio of activities perused by the rural households from cultivation, animal husbandry, casual wage works, regular wage works and other non-farm activities, etc. Income is accrued from these diverse activities as well as from transfer income such as pension and remittances. Sources of income are determined by various factors such as assets base of the households (farm and non-farm), level of education and skill attainment and employment opportunities available.

Cultivation and animal husbandry activities together contribute to about one-third (32%) of the total income. Regular employment is yet another important source of income that constitutes about 25 per cent followed by self-employment activities contributing to about 21 per cent income. Transfer income (pension) also contributes substantively (10%) and remittances accounts for about 9 per cent to the total household income. Uttarakhand hill economy has been postulated as highly dependent on remittance income and often termed as money order economy; however this line of argument does not seem to bear much empirical evidences. Income from casual income (including government programmes) is minuscule contributing 2 per cent in the share of household income (Figure 7.1).

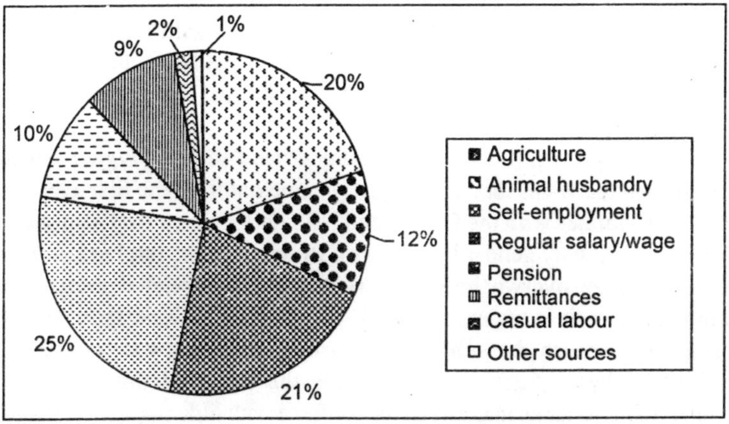

Fig. 7.1: Proportion of Income from Different Sources

Share of income varies across household features from different activities. It can be observed that the household income of Brahmins and ST is significantly higher compared to other social categories. Average annual income of 'other castes' is the lowest followed by SC (Figure 7.2).

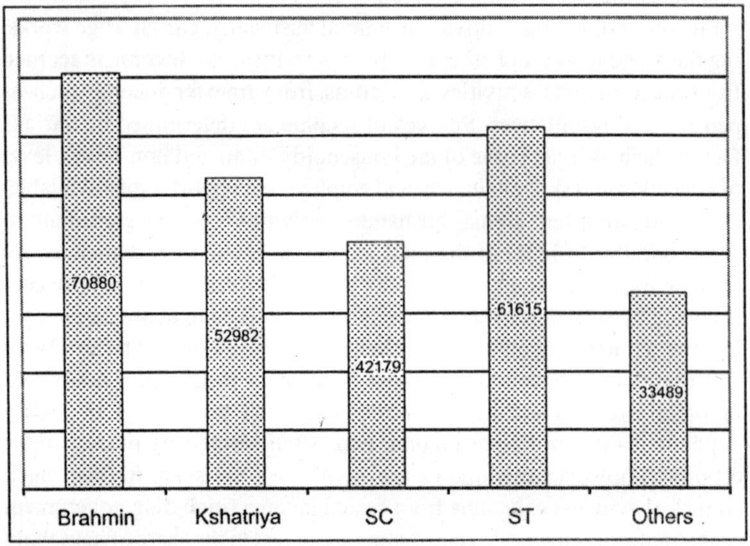

Fig. 7.2: Average Annual Incomes by Social Category

Majority of Brahmins and ST are associated with highest income quintiles, while the poorest social groups— the SC and the 'other castes'— are concentrated in the lowest income quintiles and the Kshatriya are in the middle quintiles (Table 7.5).

Typically the hill economy of Uttarakhand showing higher female LFPR than the males that has distinctly come out from the field data and also corroborates with the Census and NSS data. Survey data also shows that there is highly gender biased work structure in the rural areas of the State as women overwhelmingly work in agriculture-related occupations while their male counterparts work in non-farm occupations. This is primarily because the subsistence farming economy with precarious industrial base. Agriculture assumes the predominant sector for employment for large majority of labour force with little surplus generating capacity and by and large this sector acts as labour sponge, in particular

Table 7.5: Distribution of Household Population by Social Category and Income Quintile Groups

Per capita income quintile group	*Brahmin*	*Kshatriya*	*SC*	*ST*	*Others*
Bottom	14.55	23.14	25.93	2.00	33.33
2nd	14.55	20.09	33.33	6.00	41.67
3rd	14.55	21.40	20.37	18.00	25.00
4th	21.82	20.09	9.26	34.00	0.00
Top	34.55	15.28	11.11	40.00	0.00
Total	100.00	100.00	100.00	100.00	100.00

Source: Field Survey, 2004-05.

for the females. Non-farm activities constituting petty trade and business wage/salaried employment and casual employment particularly in construction activities have been by and large male dominated activities and female have a very low share in these activities. Generally, developed Blocks showing higher share in non-farm activities than the less developed ones primarily because of diversification from cereal crops to horticulture crops and better infrastructure and proximity to Block and district headquarters.

Agriculture is increasingly becoming an uneconomic family enterprise, and employment opportunities outside agriculture are extremely limited, particularly for females. Within farm sector there is huge incidence of underemployment in terms of unutilized labour time and majority of rural households are forced to diversify their activities as a part of their survival strategy to cope seasonality and uncertainty of production. The survey results clearly show that the problem of unemployment is huge one in the hill districts of the State and it is more pronounced among relatively less developed Blocks than the relatively developed Blocks. Unemployment is concentrated more among lower social classes and among landless class. Intensity of unemployment is visible among the higher educated categories.

Females are disproportionately represented in the agriculture and allied activities (90%) with low levels of education and a very high proportion of them are illiterates. Generally low levels of education of a large majority of workers in rural areas of Uttarakhand, particularly of females, have implications from the policy perspective for their employability outside the agriculture. Their low educational levels can hardly help them to secure employment, outside agriculture.

Average annual days of employment available in agriculture and allied activities are indeed very low in the hill areas. Although, people seem to be working very hard in agriculture activities yet it does not provide employment beyond few months in a year. On an average agriculture and allied activities provide employment to about 56 days in a year for males while for females it provides for about 106 days of employment.

However, it can be observed that the relatively developed Blocks have generally higher person day's employment compared to less developed Blocks. The reasons for higher days of employment in the relatively developed Blocks could be due to more areas under horticulture crops leading to commercialization of agriculture (vegetables, fruits and other non-food crops) compared to relatively less developed Blocks. Availability of casual employment is less than two months (50 days) in a year, providing on an average 51 days employment for males and 43 days for females.

Multiple economic activities are most pervasive and widespread among different groups of population in the survey areas in Uttarakhand, manifesting symptoms of backward economy. The average annual income of the households is estimated to be low with significant variations across different castes and land size classes. The principal reasons for variations of income across the castes are linked closely to the asset base of the households and also educational and skill levels that have overarching influence on income generating capacity. It can be observed that the relatively developed Blocks showing low proportion of their population in the low income groups and less developed Blocks showing relatively higher proportion of population in the lower income groups.

A large majority of workforce (75 per cent) is engaged in agriculture and allied activities contribute to about one-third of total household income and mean earnings from this source is lowest compared to that of self-employed and casual wages in non-farm activities that validate our hypothesis that this sector has limited potential for absorption of growing labour force productively.

Policy Implications

This chapter brings out the some of the major issues that have wider ramifications for policy formulation and implementation with a view to improving livelihoods in the mountain regions. Some of the results have direct implications for policies and programmes based on evidences drawn from conceptual framework of mountain economies and supplemented by household survey in one of the States (Uttarakhand) that have great underpinning to evolve long term policy for sustainable livelihood security and raising living standards.

(i) Diversification Strategy

Traditional hill agricultural practices have become increasingly uneconomical and, therefore, there is a need for diversifications from field crops to horticulture and non-food crops that have enormous potential for improvement in livelihoods and income generation. Area studies have also indicated that there has been substantial transfer of area from cereal production to vegetable production with greater advantages to the farmers in terms of income and employment. From the long-term perspective, this shift from low value field crops to high value added horticultural crops seems to be most obvious and strategic option in the region. However, lack of proper institutional mechanism such as credit, processing, storage, post-harvesting technologies and marketing network have severely hampered the systemic growth of this sector in Uttarakhand. If all these services are efficiently organised with necessary institutional back up, horticulture can generate significant employment and earning opportunities in the region. In certain parts of the region, particularly in lower and middle Himalayas, it offers significant scope for development and even could become a lead sector. Establishment of linkages from growers to local processing and to large plants (within and outside region) is most crucial for sustainable growth of horticulture. Enterprise based activities, based on local resources, is critical for sustainable employment generation and for enhancing income levels, which in turn can help in improving the living standard of the people. With appropriate policy interventions it is possible to develop enterprise based activities that would provide higher income through processing and value addition, and help develop upstream and downstream linkages necessary to generate internal momentum of growth. Such enterprise-based activities have been developed in certain parts of the hilly areas in the HKH region based on niche products.

Experience of Himachal Pradesh is worth emulating for development of horticulture. Our findings also support the hypothesis that higher area under horticulture helps restrict out-migration. It is, therefore, important from policy point of view to promote this segment with adequate support system.

Improving the asset base of the poor is crucial. Present research evidence has shown that poor farm households often lack assets (such as land, education and skill) that serve as important capacity variables for participating in Rural Non-Farm (RNF) activities. However, in view of limited availability of cultivable land, it becomes an extremely difficult policy option to redistribute or distribute land to those who are landless or fall in sub-marginal category. Developing appropriate education and marketable skills are, therefore, the other important policy options for accessing employment in the skilled labour market.

(ii) Development of Appropriate Educational and Skill Structure

In spite of relatively distinct resource base and potential for the development of numerous non-farm activities and enterprises in the State, its potential has not been fully exploited to the extent its resource base permits within the constraints of fragile environment. There is a need for converting these physical resources into outcomes in terms of enterprises with value addition activities, capacity building and human resource development. Transformation from the physical into real resource base would generate demand for numerous types of new skills and competencies, which would ultimately create employment and incomes in the productive activities. For this, the education system has to be made relevant and this would require revamping of existing educational system towards meeting the requirements of the emerging skill structure. Relevance implies a correspondence between the nature of skills required by the enterprises and the type of skills provided by the educational and training institutions.

In this context, vocationalisation of education is of paramount importance with a focus on emerging and market oriented trades. It is critical and so central in a rapidly changing economic environment that has profound impact on the production process, the demand pattern and the skill requirements. Such a change in economic structure requires not only huge capital investment but also a workforce that has the flexibility to acquire new skills for new jobs as the structure of economy and occupation changes.

Vocational education is not only grossly inadequate in the State but seems to be less relevant to the needs of economy. Technical courses like ITIs and Polytechnics suffer from declining intake, obsolete trades, poor infrastructure and weak industry-education linkages. Many professional courses are mushrooming, albeit in recent years, but then without corresponding to the emerging demand of economy in terms of skill and competency levels, they remain qualitatively at sub-optimal levels. Without any evaluation and impact assessment on regular basis, its efficacy to respond the needs of changing economic environment remains largely unknown. Such a situation might result in low returns, which would not be a viable form of investment in the long run, and this kind of situation needs to be corrected by setting some standards and benchmarking with a view to improving the quality of these institutions.

Mismatch is largely due to lack of relevance of course curriculum and inflexibility in education system that does not respond to the changing economic environment. Unless these technical and vocational courses are qualitatively improved to make them marketable, they would continue to become less relevant to the needs of the market.

The policy message seems fairly straightforward. Efforts to improve relevant marketable skills are important in an effort to promote employment into high income and the importance of enhanced rural skill base for development of the rural non-farm (RNF) sector is incontrovertible. Our regression results clearly reveal that education and skill are strong determinants for enhancing household income.

(iii) Micro-enterprises and its Linkages

A substantively large majority of enterprises appear to be dominated by trade and household manufacturing activities with narrow profile of activities. This kind of enterprise-based activity has limited product lines, carried out on a small scale with little surplus for reinvestment. There are sizeable constraints in terms of scale and efficiency that result in weak linkage effects. Entrepreneurial activities suffer from numerous internal and external constraints. But over the years, these enterprises are languishing and operating at a low equilibrium trap with little product diversification and surplus. The major reasons for sickness of these enterprises can be attributed to both the supply and demand side aspects.

Rural areas appear to be poorly equipped in terms of economic (roads, electricity, communication, etc.) and social infrastructures (vocational and training institutions, market information, access to credit institutions, etc.). Some of these constraints are typically associated with hill specificities (inaccessibility and fragility, for example) that pose constrain to spur entrepreneurial activities.

Family labour is more prominent in the rural enterprises with minuscule hired labour. Access to various supply side inputs (e.g., credit, technology and market) is low and most of the output of these enterprises is meant for the local markets. Demand is low primarily because of extremely low incomes and purchasing power of those who work. In any case, this is an inadequate base for growth. But, if demand for its output is to grow significantly, it is necessary to strengthen market linkages. Input linkages are extremely weak to the local economy that generate little or no value addition to the enterprises.

It is argued here that mountain areas should primarily focus on niche products in which the region has comparative advantage rather than producing those products that does not have distinct advantage in terms of product and price competitiveness. However, there is a need for evolving special policies and support mechanisms to promote the micro and small enterprises taking into account the diversity and constraints in the regions.

Adequate employment opportunities cannot be generated unless productive base of economy is enhanced. This would require sector-specific development strategies integrating environmental, economic and social components of sustainable mountain development. For this, detailed planning is a prerequisite in the region and within regions due to enormous diversity in terms of suitability of land and other resources. In fact, land capability differs considerably according to altitude and slope of the hills, therefore there is a need for appropriate land-use patterns across regions and sub-regions.

It is, therefore, necessary to formulate mountain perspective framework that incorporates growth with equity and that addresses the issues like mountain agriculture, persistent poverty, sustainable livelihood, environment sustainability and gender dimension. Such perspectives would call for an integrated approach to development dovetailing the mountain specificities adequately in order to achieve the desired objectives.

REFERENCES

Awasthi, I.C. (2012). *Livelihood Diversities in Mountain Economy: Constraints and Opportunities*, Concept Publishing Company, Delhi.

—— R.P. Mamgain and Balwant Singh (2005). "Employment in Uttaranchal: Issues and Challenges", *The Indian Economic Journal*, Volume 53, Number 2, July-September.

Choudhury, Dhrupad (2004). "Agricultural Transformations in the Uplands of Northeast India: Where Are We Heading?", paper presented in a National Seminar *Transforming the Hill Economy of North-East Region: Emerging Challenges and Opportunities,* November 26-28, 2004, Itanagar.

Das, Gurudas and Rajesh Datta (2004). "Structural Transformation of the Hill Economies of India's North East: Some Policy Implications", paper presented in a National Seminar on *Transforming the Hill Economy of North-East Region: Emerging Challenges and Opportunities,* November 26-28, 2004, Itanagar.

Davies, S. (1996). *Adaptable Livelihoods: Coping with Food Insecurity in the Malian Sahel,* Macmillan Press, London.

Ellis, Frank (2000). *Rural Livelihoods and Diversity in Developing Countries,* Oxford University Press, New York.

Gordon, Ann and Catherine Craig (2001). *Rural Non-farm Activities and Poverty Alleviation in Sub-Saharan Africa, Centre for Sustainable Development,* Natural Resource Institute, DFID, Policy Series 14, UK.

Goswami, Homeswar and Jayanta Kr. Gogoi (2004). "Regional Disparities in India with Special Reference to Hill States of Northeast India", paper presented in a National Seminar on *Transforming the Hill Economy of North-East Region: Emerging Challenges and Opportunities,* November 26-28, 2004, Itanagar.

Maithani, B.P. (2004). "Land Policy & Land Management: Key to Transformation of Northeastern Hill Economy', paper presented in a National Seminar on *Transforming the Hill Economy of North-East Region: Emerging Challenges and Opportunities*, November 26-28, 2004, Itanagar.

Negi, Aparna and Pyare Lal Verma (2004). "Himachal Pradesh: A Model for Hill Development", paper presented in a National Seminar on *Transforming the Hill Economy of North-East Region: Emerging Challenges and Opportunities*, November 26-28, 2004, Itanagar.

Papola,T.S. (2003). "Development and Livelihoods in Sikkim: Towards a Comparative Advantage Based Strategy", paper prepared for *UNDP*, New Delhi (mimeo).

Planning Commission (2002). *Tenth Five-Year Plan (2002-2007).* Chapter 4, Special Area Programmes, Government of India.

Pokhriyal, H. C. (2004). "Mountain Development: Lesson Learnt and Strategic Initiatives", paper presented in a National Seminar on *Transforming the Hill Economy of North-East Region: Emerging Challenges and Opportunities*, November 26-28, 2004, Itanagar.

Sadhu, A. N. and Falendra K. Sudan (2004). "Experiences of Jammu and Kashmir Hill Economy and its Lessons for North-East", paper presented in a National Seminar on *Transforming the Hill Economy of North-East Region: Emerging Challenges and Opportunities*, November 26-28, Itanagar.

Stark, Oded (1991). *The Migration of Labour*, Harvard University Press, Cambridge, MA.

8

Towards Improving Viability of Small Farmers

Brajesh Jha

Introduction

According to Agriculture Census 2005-06 out of 1.29 lakh total agriculture holdings, 19 per cent are small and 65 per cent are marginal farmers in India. If we further categorize marginal farmers, the farmers with landholdings between 0.01 to 0.4 hectares accounts for around 40 per cent of total agriculture holdings in the country. This group has been referred here as sub-marginal farmers. Nevertheless a glance to different issues of Agriculture Census shows that number of small and marginal holdings between 1971-91 increased by around 50 and 75 per cent; while that of the large farmers has decreased by 40 per cent during the same period. After 1991 all operational holdings barring marginal have started decreasing in the country. These statistics suggest that marginal and sub-marginal famers in the country will continue to increase in number. Bulk of these holdings are not viable on their own. Farmers lacking sufficient off-farm income opportunities in their vicinity leave their land under some sub-optimal arrangements, and migrate for their livelihood. Often they work in unorganized sector, live in unhygienic conditions and turn up as urban rather rural poor.

The marginal and small farmers now account for around half of the operated area under agriculture. The viability of these farmers is, therefore, important not only for an average farmer's subsistence, but also for the robust growth of agriculture in the country. Since non-agriculture sector is not absorbing the growing work force in the country, pressure on land for livelihood would continue and small and marginal farms would dominate. The small farms also have certain benefits. The small farms are more sustainable as intensity of livestock per unit of land is higher on these farms; these farms are also more integrated. Such integrated farms not only enrich soil health but also improve nutrition security of farm households. The small farm is also a kind of own account of enterprises, a large number of small independent enterprises are good for the health of a democratic society. The NSS result on employment 2004-05 however shows that more than one-third of rural poor are these farmers. These farmers adopt multiple activities for their livelihood. Therefore, there are enough reasons to improve the livelihood of these farmers by making them viable. In view of this it is important to assess viability of small farmers and look into ways of improving viability of small farmers in India. The next sections of the paper illustrate on the above issues.

Viability of Small Farm

There are numerous studies to suggest that small farms in a large part of country are not viable (Jha D., 2001, Singla and Sidhu 2010). There is dearth of a pan-Indian studies in this regard; the National Sample Survey (NSS) results on Situation Assessment of Farmers (SAF), 2005 (NSSO 2005), however, are important.

Table 8.1: Average Monthly Income and Expenditure in thousands of Farm Households of different Land Size Classes

Land size Classes in hectare	*Average Monthly*	
	Income (Rs.)	*Expenditure (Rs.)*
Less than 0.01	1380	2297
0.01 to 0.4	1633	2390
0.41 to 1.0	1809	2672
1.01 to 2.0	2493	3148
2.01 to 4.0	3589	3685
4.01 to 10.0	5681	4626

Source: NSS Report No. 497.

Table 8.1 presents average monthly income and expenditure of farm households of different land size classes. It depicts that the average monthly expenditure of farm households is significantly higher than average monthly income of households of farmers less than 4 hectares. Similar results are reported by other studies as well. Yet merits of the above results cannot be overemphasized as the NSS (2005) is based on samples of farmers drawn from different parts of India.

Reasons for Low Profitability on Small Farm

There are many reasons for non-viability of small farmers. In fact, there are may factors that directly / indirectly affect profitability of small farms. These factors / constraints can broadly be grouped as market- technology-resource-related constraints. In market related constraints unequal exchange of small farmers with buyer of agri-output market and seller of agri-input market is well documented. The marketing disadvantage for small farmers is also on account of high transaction cost since small farmers have low marketed surplus. The small farmers have lower capability to withstand adverse situations; they are more prone to external pressures. These pressures can be in the form of depletion of production following changes in climatic factors. The external pressures can also be in the nature of lower price of producer following relatively better integration of domestic and world markets.

Though biochemical technology is scale neutral, there are many other technologies highly biased against size of land. For the present discussion technologies have been grouped as mechanical, natural resource management related and knowledge-intensive technologies. Many of the knowledge-based technologies are accessed through radio, cell phone and TV. Bulk of small and marginal farmers has radio and cell phones only. In addition, knowledge-based technology also requires certain level of education. The natural resource management technology has wide ranges of technological options; it ranges from drip to submersible technologies. While drip is relatively scale neutral; submersible technology is not only scale biased but has also emerged as an important source of interpersonal inequity in the large parts of India (Jha, 2000). The mechanical technology is highly scale biased. This has been so from beginning but particular relation has further deteriorated with the high cost of labour in rural India following MNREGS in the country.

At times technology generating institutions are also biased against small as they bring limited technology for small farmers/ producers. The knowledge disseminating institutions especially extension agencies also ignore small farmers as their target is mostly to achieve certain amount of production/ productivity from the field. For a large group of marginal farmers, question of viability remain important, in spite of the adoption of certain technologies. Mixed farming has, therefore, been the way of life of Indian farmers. These incomes are often not sufficient especially for marginal and sub-marginal farmers. They have to depend on wage income in either agriculture or non-agriculture sector. Farmer's income at times is also generated from self-employment in their own non-agriculture enterprises.

Multiple Sources of Household Income of Farm Households

The NSS Situation Assessment of Farmers, 2005, provides information about the sources of income of farmers. Fig 8.1 presents farm households of different land sizes on the basis of income sources. These sources of incomes are wages, as obtained from casual work, income/receipts available separately from cultivation of crops, farming of animals, and non-farming business. In Fig. 8.1, farmers, on the basis of size class of land, are classified as large (more than 10 hectares), medium (4-10 hectares), semi-medium (2-4 hectares), small (1-2 hectares), and marginal (less than one hectares). With the increased pressure on land, marginal farmers number is substantially increasing. These farmers are particularly important for studying the dependence of farmers on off-farm income. The NSS 2005 report further categorizes marginal farmers of less than one hectare into three groups: 0.41-1.00 hectare, 0.01-0.40 hectare and less than 0.01 hectare. Figure 8.1 presents contribution of different sources

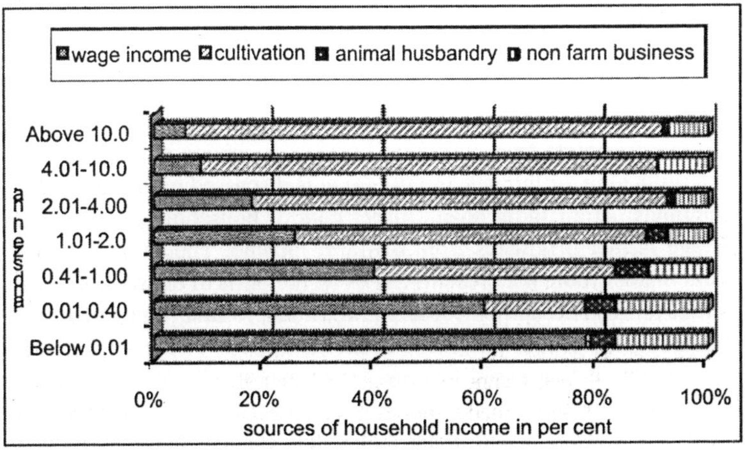

Fig. 8.1: Sources of Farmers Income by Farm Size.
Source: Author's computation based on NSS Report No. 497.

of income of the farm households of different land size classes. It shows that as the size of landholding decreases, contribution of crop husbandry decreases and income from wage and non-farm business increases. Farming of animals is an important source of income for small and marginal farmers.

The contribution of farm animals in household income decreases with the increase of size class of land. This contribution declined significantly on the medium and large farms. It is widely recognized that in India rearing of cattle on small farms is profitable, largely with the surplus family labour. As the size of landholding increases, surplus family labour available for cattle rearing decreases; therefore, intensity of farm animals on medium and large farms is low. Farmers with the marginal holdings, in fact, earn a significant proportion of their household income from wages. The contribution of non-farm business in household income decreases with the increase of land size. But for medium and large category of farmers the share of non-farm business in household income has suddenly increased. This is not in harmony with the earlier trend and argument. One may note that the increase in non-farm income of farmers in the medium and large category is possibly a result of development-induced rural diversification; while non-farm business for lower size class of holdings is possibly a distress-induced diversification. Examples of development-induced rural diversification are employment in relatively high wage sector. Employment in such sector requires certain level of education and similar other enabling conditions; ironically medium and large farmers only afford to enable their children with the above capabilities.

The casual work for wages can be in agriculture or non-agricultural activities, but it would definitely not be on the own farm. Income from wages and receipts from non-farm businesses are together grouped as off-farm income of farmers and the same is discussed in the subsequent paragraphs. At all India level, 50 per cent of household income of an average farmer originates from off-farm income. The situation varies across States. Table 8.2 presents States on the basis of dependence of an average farm on off-farm income. In Rajasthan, Kerala, Tamil Nadu, and West Bengal, off-farm sources account for 75, 68, 64 and 61 per cent, respectively, of household income for an average farmer; whereas, in States like Meghalaya, Uttaranchal, Punjab, and Bihar, off-farm income accounts for 25, 31, 38 and 39 per cent of household income, respectively. The above examples suggest that resource endowment: natural or acquired, distribution of land, status of non-agricultural sector in a State, level of urbanisation and similar other factors are responsible for differential share of off-farm income for farm households in different States of India.

Table 8.2: Dependence of an Average Farm Household on Off-Farm Income across States

States with values in per cent in parentheses
Arunachal Pradesh (71.9), Chhattisgarh (50.1), Haryana (56.4), Himachal Pradesh (63.6), Jharkhand (54.7), Kerala (68.2), Manipur (64.3), Orissa (66.8), Rajasthan (75.7), Tamil Nadu (62.9), West Bengal (60.8)
Andhra Pradesh (48.8), Assam (38.8), Bihar (38.6), Gujarat (39.7), Jammu and Kashmir (48.8), Karnataka (46.6), Madhya Pradesh (46.2), Maharashtra (42.8), Meghalaya (25.2), Mizoram (35.7), Nagaland (47.5), Punjab (38.3), Sikkim (46.2), Tripura (46.3), Uttar Pradesh (45.5), Uttaranchal (31.3)

Note: Off Farm income consists of income from wages and non-farm business of farm households as reported in NSS Report No. 497: Income Expenditure and Productive Assets of Farm Households, 2005.

In brief, the above discussion illustrates pattern of employment with different sets of data. The NSS data on employment shows a sharp fall in the share of agriculture in total rural workforce in the recent period (1999-00 and 2004-05). Rural employment during this period has been propelled by construction, trade and transport; in rural India, employment in trade has emerged as a residual. Though manufacturing is the most important industry in the non-farm sector, low productivity of manufacturing in the rural sector leaves sufficient scope of its further improvement (Table 8.2). Women now account for around 30 per cent of the workforce in rural India, and a bulk of them (85%) are working in agriculture. Other industries that account for a significant proportion of female workers are manufacturing, community services, trade and construction. Interestingly, share of female workers is increasing in the former and decreasing in the latter group of industries.

Policy for Improving Off-Farm Income and Viability of Small Farmers

The discussions in the above sections suggest that small farmers depend on multiplicity of activities for their livelihood, yet majority of these farmers are poor. This essentially warrants increase in income from different constituents of household income of farmers. The small holding agriculture provides limited potential of increasing income from agriculture, though certain high value crops like apple, rubber are exceptions. Certain land saving enterprises like livestock rearing also provides options, but increasingly complementary relationship between crop and livestock enterprises are weakening. All these warrant contribution of sectors other than agriculture in the household income of farmers. There are two different ways of increasing the contribution of off-farm income. First, it is by migration of family members in search of livelihood and increased remittances of farm family. Second, it is by increasing off-farm income of farmers. The off-farm income of farmers arises from sectors other than agriculture in the rural vicinity. This sector is loosely referred as rural non farm sector. The Mahatma Gandhi National Rural Employment Guarantee Act (MGNREGA) is also helping farmers in earning off farm income, but accepting such programme as part of long term development policy is difficult, especially when it is reported to have been affecting production of other sectors of the rural economy.

Though one of the above sources of income is important to make an average farm household viable, the role of remittance in the household income of an average farmer is on the rise. The remittance is an off-shoot of migration. It is however leading to several kinds of problems like, feminization of agriculture and, absentee landlords in the villages. Either of these is not good for the growth of agriculture in the country. The migration of rural workers to urban centres is beset with other sets of problems like further deterioration of civic amenities / facilities, inhuman living conditions of workers in urban sector. Unless there is more explicit industrial and urban policy that accommodate these workers and provide a relatively decent life in unban vicinity, the city-centric growth of remittance for farm households has its limits and are often not welfare enhancing. Its ill effect on agriculture is underestimated. The present study, therefore, argues for off-farm income in rural vicinity. This has numerous advantages. The next sub-section illustrates policy that will improve viability of farmers by increasing off farm income and strengthening social security measures. A robust growth in off-farm income also requires growth in agriculture and similar other sectors of rural economy. The policy recommendations will largely draw from experiences of policies and programmes in rural India. The same is discussed in brief under the following four headings:

- Macroeconomic Policies;

- Sectoral Policies;
- Institutions and Incentives for Rural Employment; and
- Active Labour Market Policies.

Macroeconomic Policies with Important Implications for Rural Sector

Historically there have been several fiscal and monetary incentives that may have played a role in encouraging an excessive use of capital relative to labour in rural sector. Some of these measures were originally conceived as instrument for encouraging investments and productivity of industries. An indirect effect of such measures has, however, been to distort the relative prices of capital *vis-à-vis* labour (critical factors of production) in the region. Review of economic policies suggests that several economic policies have encouraged use of capital, while labour is the most abundant factors of production in India. The subsidy for tractor during the earlier decades of green revolution is an example of the above policy distortion in the country. One possible way of correcting the bias against labour intensive production is to subsidize labour intensive rural enterprises. This can be in different forms. Investment subsidy on condition of realizing a targeted level of employment per unit of investment is one example. This may encourage concentration of certain labour intensive industries in the rural areas.

The importance of trade policy has increased with the globalization of economy. The opening up of trade for agriculture commodities has been instrumental in increasing cropped area under high value horticulture commodities. Similarly, reduction of import duties in manufactured goods has increased import of manufactured commodities from the low cost countries like China. This has adversely affected production of many commodities important for rural sector. The opening of economy has also made the policy of Reservation of Industries for Small Scale Sector obsolete. A wider distribution of small scale industries to some extent is essential for encouraging growth of rural non-farm sector in the country.

Fiscal policy has also been instrumental in transferring resources to rural sectors. (There are many government programmes and policies to address concerns of rural economy. Some of these programmes have implications for increasing viability of small farmers. Most of these programmes, however, suffer from one or other implementation related problems. Efficiency of public expenditure on these programmes and the growing wedge between outlay and outcome of programmes are matters of concern. Some of these concerns can be addressed at the micro level through the district plans. A large amount of resources for the above programmes can be released with the convergence of various development programmes in the district plan. The chances of such convergence are not remote with an efficient district plan involving Panchayati Raj institution.

Policies related to Specific Sectors Important for Rural India

As discussed earlier viability of small and marginal farmers require off farm income; the same requires robust growth in rural non-farm sector. Rural non-farm employment includes several heterogeneous non-farm activities that have different demand and supply conditions in their respective input and output markets. The industry-specific policy is, therefore, important for the growth of rural non-farm employment in the country. According to the CSO classification there are nine industries. Only policies related to agriculture, manufacturing, business and services in particular are discussed here in brief in the following paragraphs.

Agriculture

Government policy that has potential to increase factor productivity would increase viability of a farm. In this regard rationalization of farm input prices is important. Once chemical input reflect its actual price (shadow price), farmers will be encouraged to adopt various resource conserving technologies that may promote low-external input sustainable agriculture in the country. Such policy may encourage use of bio-fertilizers and bio-pesticides. Many of these bio-fertilizers can be produced on small farms and are also labour-intensive. Rationalization of farm output price may encourage shift from crop to horticulture and plantation crops. The same is reported to have been yielding higher value of output per unit of land. Profitability or return per unit of land in agriculture will further increase by the increased vertical integration of farm with the processing units and markets. Suitable infrastructure and institutions that reduce inefficiency in the post-harvest operations will increase productivity per unit of land.

Indian agriculture is changing in many ways; besides smaller size of landholding, agriculture is becoming knowledge intensive and also market responsive. In this context, persons who can disseminate knowledge on both production and marketing aspects of agricultural commodities are desired in the country. Further with the fragmentation of landholding need for service providers of lumpy goods like tractors have increased in the rural economy. Feminization in agriculture has also increased but there are evidences of disguised and seasonal unemployment of female workers in rural India. Therefore rural enterprises based on primary resources are desired for females. Examples of such enterprises are herbal medicines, cosmetics, tissue culture, floriculture, agro-processing at primary level, non-conventional energy sources, extraction of vegetable dyes. The lesser exploited allied sector activities like apiculture, sericulture, rearing of birds and small ruminants may also be encouraged on small farms largely dominated by females.

Allied Sub-Sector

Livestock sub-sector is growing faster than many other sub-sectors of agriculture. A regime of favourable price accompanied with the desired changes in the structure of livestock population has brought about the above changes in the growth of the livestock economy. Livestock-based rural livellhoods have emerged important with the increased fragmentation of land and increased number of small and marginal farmers. The demand side impetus for the growth of the livestock sector remains strong. Constraints are appearing on the supply side of milk and milk products and this is on several accounts. The decline and degradation of common pool resources is one. Much of the growth in allied sector is following the complementary relationship between crop and livestock sector. However, the complementary relationship between these sectors is weakening with the urbanization-induced growth of dairy sector. In fact competitive relationship between crop and livestock sector is strengthening on several accounts (for details see Jha *et al.*, 2009). There is need to reinforce complementary relationship between crop and dairy enterprises by providing better price of milk to farmers in rural hinterland. Government effort to create infrastructure or institution that will increase farmers share in the consumers' spending for milk and milk products remains important. A better price of milk for farmers may encourage farmers to adopt proper breeding, feeding and management practices.

Industry, Business and Services

Industry, business and services in rural sector are loosely referred as Rural non-farm sector. This is the prime source of off-farm income of farmers.

In development literature two broad strategies to increase rural non-farm employment are: first, the agriculture led development paradigm presumes that with the increase of productivity in agriculture, manufacturing and other sectors of the rural economy would take off. Second, the urban and export led growth of rural non-farm employment presumes that as structural transformation proceeds, share of agriculture in rural employment and income declines; and that of other sectors in rural economy increases in the country. The second strategy is becoming important in the recent period following rural transformation with adoption of the above strategy in some of the East Asian Countries. In these countries manufacturing has emerged important following international demand for cheap products and further integration of their manufacturing units with the rural hinterland. Incidentally service-led growth in India has by and large excluded rural services. In a diverse country like India, strategies for increasing productive employment in the rural sector must vary across regions. The present study, while recognizing pivotal position of agriculture in rural transformation, suggests for a manufacturing and tourism led transformation of rural non-farm sector in the country.

The village resource-based organic manufacturing, the labour intensive manufacturing, skill intensive rural artisans' work are some of the typologies of manufacturing activities for rural sector. Table 8.3 enlists some of the manufacturing activities emerged important for rural sector in the latter group of manufacturing. Again in rural sector smaller enterprises more specifically the own account enterprises (OAEs) dominate, therefore, policies related to micro and small enterprises are important. In the initial years of planned development, village and small scale industries were largely perceived to fulfil the demand of consumer goods. Such industries were promoted with the help of Khadi and Village Industries Corporation (KVIC). Subsequently, many of the consumer goods were reserved for production in the small scale industries. To protect these industries from each other and from the organized sector, a cluster approach for development of small industries was adopted. Today, the policy of reservation has, however, become obsolete in an open economy. Therefore, the Micro, Small and Medium Enterprises Development (MSMED) Act came in force since 2006. The MSMED Act is a supportive policy and is a significant departure from protective policies of the past; and it aims to facilitate growth of similar kinds of enterprises in the country. Though the MSMED Act encompasses manufacturing, business and service based enterprises. The Act has an urban bias and in this context, the initiative of the State government of Rajasthan in creating rural non-farm development agencies (RUDA) to promote rural non-farm enterprises is important (Jha, 2011).

Table 8.3: Emerging Rural Man—Manufacturing Activities

Manufacture of wooden implements, furniture
Manufacture of bamboo and cane articles
Manufacture of leather footwear, objects
Manufacture of carpets, rugs and other coverings
Embroidery and embossing of leather articles
Preparation of dolls and toys
Manufacture of matches
Potteries and clay making
Wood carving, stone carving
Folk paintings on textiles, etc,
Processing of Agri-based products

Rural tourism is an easy way of injecting income into the rural economy. In certain regions of the country, rural tourism may also emerge as the engine of rural non-farm growth. Though tourism in India is generally associated with the historical antecedents, scenes and beauties of nature of a region (example hills, sea); there is latent demand for another kind of tourism: cultural tourism, orchard and horticulture-based tourism.

The study, proposes that a district may be identified with certain manufacturing or tourism activities; and desired facilities for these activities may be developed according to the strength and resource endowment of the area. The industries that are closely integrated with the primary resources of the region may be identified. Such identification after proper industrial potential survey of the region is important. With the increased importance of decentralization, a district-level planning and identifications of industries for a district are emerging reality. Such identification of rural industries must start with the District Industry Centre. The creation of industry at the level of district besides others also requires land but in the recent years there are frequent problems in acquisition of land for industrial purposes in the country. Incidentally many of the industrial estates created by government remain underutilized in dearth of key infrastructure. Nevertheless, many farmers also want to come out of agriculture as bulk of marginal and sub-marginal farms are not viable in the large part of country (NSS, SAF, 2005).

The study, therefore, proposes that the land need of industry can be met with the pooling and consolidation of land obtained from these farmers. All this requires different kinds of coordination among various stakeholders: farmers, development policy formulators and implementers in government, development practitioners in non-government organizations and industrialists at the district level. The present study also argues for setting up of a apolitical statutory body in the State to look into any acquisition of land beyond a certain size (250 hectares). The statutory body may be assigned the responsibility of pricing of land, compensation, resettlement and employment of the affected people. The land price must not ignore the long run cost of land especially the quality of land. The cost of land at the same time should not be too high to discourage industrialists from rural industrialization. To avoid such situation some successful examples of farmer based companies also need to be replicated in different parts of the country[3].

The sectoral policies in brief argue that growth in agriculture, manufacturing and tourism would trigger growth of productive employment in other sectors of the rural non-farm economy like utilities, trade, business and services. Manufacturing, in particular, is important for a broad-based growth of productive employment in rural India. Though there have been numerous public institutions to encourage rural manufacturing, these institutions have failed to reduce the burgeoning gap between the productivity of rural and urban manufacturing (Table 8.4). This also failed to check the flight of manufacturing enterprises away from the rural sector. The present study, therefore, argues for sufficient incentives to encourage manufacturing in the rural vicinity.

Table 8.4: Gross Value Added (GVA) Per Worker in Unorganised Manufacturing (in '000 Rs., 2005-06)

Enterprises	*Rural*	*Urban*	*All*
OAME	10.4	16.3	11.3
NDME	26.9	43.4	36.5
DME	41.9	65.3	55.1
All	16.2	38.2.	24.0

Source: NSS Report No. 526.

Institutions and Incentives for Growth of Productive Employment in Rural Sector

In spite of the increasing importance of private sector in the neo-liberal framework the role of the State remains important. Certain government policies, though basic for the growth of industry, are not industry specific (Examples under this category include technology generation, skill formation, and infrastructure creation.) and there are frequent cases of 'market failure' in provisioning and access to such facilities, due to existence of 'externalities' and 'free riders' in the industry. The role of State in increasing capabilities of workers by providing better education and health facilities cannot be overemphasized. Nevertheless, since bulk of farmers are small, poor and are in the unorganized sector without a reliable safety net, the role of State in protecting such farmers from contingencies such as illness, accident, untimely death of bread-winner and old age becomes important. Though Union Government's expenditure on some of the above accounts has increased significantly in the recent years, still these are largely perceived inadequate. There is need to involve private sectors, voluntary organizations, interest groups in extending many of the above facilities. These alternatives are emerging important, but Government support at times is important to encourage them to remain in business. At times different kind of hybrids of institutions also emerge; the self help group as new form of cooperatives have also emerged as an important institution for development programmes. Both Government and voluntary organizations have played important role in its spread to different parts of the country. Such institutions of farmers are extremely important for intermediation of farmers with market.

Besides the above, government has also created numerous institutions to promote sector specific growth in the country. The rural non-farm sector consists of heterogeneous industries that require different kinds of institutions to regulate and facilitate an industry. The growth of rural non-farm sector often suffers in dearth of coordination between different institutions and sectors of rural economy. In this context Government of Rajasthan initiated Rural Non-Farm Development Agency (RUDA) in 1995 to promote Rural Non-Farm Enterprises (RNFEs). The agency like RUDA promotes integrated development of industry, business and services in rural sector; development of such institutions is important for the growth of off-farm income of farmers in the country.

The implementation of developmental programmes has been far from satisfactory. Inadequate understanding of local situation, lack of participation of targeted groups in development process, etc., are often cited as important reasons for the above. *The Panchayati Raj* institution (PRI), a bottom-up approach in development planning is often believed to address many of the above concerns. It is being strengthened in the successive Five Year Plan. Several multilateral organizations are also encouraging participation of alternate institutions for rural development. The Right to Information Act 2004, and provision of social audit in certain developmental programmes like MNREGS attempt to further improve implementation of programmes by encouraging transparency in government programmes at the bottom. Voluntary organizations are also important in strengthening participatory development. The National Policy on Voluntary Sector (2007) envisages government collaboration with voluntary organizations in poverty alleviation, skill promotion and entrepreneurship development.

Direct Employment Generation Programme or Active Labour Market Policies

Growth has limited trickle-down effect for certain sections of society generally left out of the developmental process (Lipton, 1983). The direct employment generation programmes (EGPs) have, therefore, become part of our development planning in the last few decades. The importance of such programmes further increases with the high incidence of seasonal and disguised unemployment of small and marginal farmers in rural economy. The employment generation programmes largely fall under two broad categories; self-employment generation and wage-based employment generation programmes. The wage-based programmes provide supplementary employment to stave-off seasonal unemployment. MNREGS is an example in the latter group which guarantees employment of one person in a household for a minimum of 100 days on asset creating public works. Such programme at distress wage has least implications for other sectors of the economy; but increase in expenditure on such programmes is often associated with a decrease of public expenditure in the social sector (Mahendra Dev, 2000).

Conclusions

The present study argues that off-farm income is important for viability of small and marginal farmers in India and their viability is also important for growth of agriculture in the country since they account for more than 40 per cent of land cultivated in agriculture. It is important that at least one sector out of agriculture, manufacturing and tourism grows for adequate growth in off-farm income opportunities. The growth in these sectors will trigger growth in other industries in rural sector. In rural manufacturing, micro and small enterprises (MSEs) play important role. These enterprises however lag behind in terms of productivity. In order to increase productivity of MSEs, they need to be linked adequately with the large and medium enterprises at the apex. The large and medium enterprises can be attracted in the rural vicinity with suitable incentives. This can be ensured in terms of infrastructure like creation of rural industrial estates, common facility centres, subsidized raw materials, etc. The non-farm sector encompasses a large number of diverse activities, success of which in a country as diverse as India requires need-based innovations in rural institutions; the RUDA in Rajasthan provides an example. Agriculture and manufacturing-based rural transformation also requires emphasis on skills, and various other services that enable farmers to earn their off-farm income and sustain livelihood. For the time being direct employment generation programme like MNREGS under distress wage remains important.

REFERENCES

Jha, Brajesh (2011). *Policies for Increasing Non-Farm Employment for Farm Households in India*, IEG Working Paper No. 310.

——, Nitesh Kumar anf Biswajit Mohanty (2009). *Pattern of Agricultural Diversification in India*, IEG Working Paper No. 302.

—— (2008). "On Stagnating Employment in Indian Agriculture", *Manpower Journal*, 43(2): 1-30.

—— (2007a). "Agricultural Wages in India – A State Level Analysis", *Indian Journal of Labour Economics,* 50(2): 293-302.

——(2007b). "Is the Role of Agriculture in Rural Diversification Decreasing", *Indian Journal of Labour Economics,* 50(3): 633-642.

——(2000). "Implications of Intensive Agriculture on Soil and Groundwater resources." *Indian Journal of Agricultural Economics*, 55(2): 182-193.

Jha, Dayanatha (2001). "Agricultural Research and Small Farm", *Indian Journal of Agricultural Economics*, 56(1):1-23.

Lewis, W. A., (1954). "Economic Development with Unlimited Supplies of Labour", *The Manchester School*, Vol. 22, No. 2, pp. 242-258.

Lipton, M. (1983). "*Poverty, Undernutrition and Hunger*", World Bank Staff Working Paper No. 597, Washington: World Bank.

Mahendra Dev, S. (2000). "Economic Reforms, Poverty, Income Distribution and Employment", *Economic and Political Weekly*, Vol. 35, No. 10 (March 4).

Mellor, John W. (1978). *The New Economics of Growth – A Strategy for India and the Developing World,* A Twentieth Century Fund Study, Cornell University Press, Ithaca.

National Sample Survey Organization(NSSO 2003). *Income Expenditure and Productive Assets of Farm Households 2003,* NSS Report No. 497, New Delhi.

———————— (NSSO 2006). *Employment and Unemployment Situation in India, 2004-05, NSS 61st Round (July 2004-June 2005).* NSS Report No. 515, Key Results, New Delhi.

NCEUS (2007). "Conditions of Work and Promotion of Livelihood in the Unorganized Sector", National Commission for Enterprises in the Unorganized Sector, GOI, New Delhi.

Papola, T.S. (1987). "Rural Industrialization and Agricultural Growth: A Case Study in India", in Islam Rizwanul (eds.) *Rural Industrialization and Employment in Asia,* ILO / ARTEP, New Delhi.

Siti Hadijah Che-Mat and Roslan Abdul-Hakim (2011). Does Farmer's Diversification into Non-Farm Employment Reduce their Likelihood of Poverty? Evidence from Malaysia. *Journal of Business and Policy Research*, 6 (1):145-155

Singh, Sukhpal (2011). Contract Farming for Agricultural Development in India: A Small Holders Perspective. Paper presented in Workshop on Policy Options and Investment Priorities for Accelerating Agricultural Productivity and Development In India on November 10-11, 2011 at India International Centre, New Delhi.

Pasha, Syed Ajmai (1991). Sustainability and Viability of Small and Marginal Farmers Animal Husbandry and Common Property Resources. *Economic and Political Weekly,* 26 (13): A27-A30.

Singla, Naresh and J.S. Sidhu (2010). Factors affecting the viability of small and marginal farmers in Punjab, *Journal of Agricultural Development and Policy*, Vol. 20 (2), pp. 51-58.

ANNEXURE 8.1

Annexure Table 8.1

Per cent Distribution of Households by Household Type for each Household Monthly per-capita Expenditure Classes in Rural India 2004-5

Expenditure class (Rs.)	*Self–employed in*		*Agricultural labour*	*Other Labour*	*Other hhs (incl. n.r.)*
	Agriculture	*Non-agri.*			
Less than 235	19.2	8.5	45.0	10.2	17.1
235-270	26.3	10.5	47.2	10.2	5.8
270-320	25.4	13.4	43.8	11.3	6.0
320-365	28.8	14.7	37.5	12.3	6.7
365-410	33.5	15.3	33.1	11.8	6.4
410-455	35.0	16.0	29.5	13.1	6.3
455-510	38.1	15.3	27.3	11.6	7.6
510-580	39.2	17.2	23.9	10.8	8.8
580-690	41.8	17.7	19.5	9.8	11.1
690-890	42.8	16.7	14.6	10.4	15.5
890-1155	42.4	17.4	9.3	9.1	21.9
1155 & above	33.7	16.9	4.2	8.1	37.2
Average MPCE (Rs.)	573	576	424	512	645

9

Small Farms and Economic Viability in Uttar Pradesh

Vidyarthi Vikas

Introduction

Productivity was bound to grow slower in agriculture than non-agriculture sector because of greater impediments to specialization and the division of labour in agriculture production. In India, marginalization of holdings is a general feature of the size distribution of operational holdings that is the percentage of holdings decreases as the holding size increases. The percentage distributions of operational holdings decline is getting progressively steeper with each decade. Large, medium and semi-medium holdings' percentages have been declining steadily since 1960-61. It is steepest for large holdings from 0.7 per cent in 1970-71 to 0.1 per cent in 2002-03 and consequently medium and large holdings declined from 6.9 per cent to 1.8 per cent during the same period. At the other end, one witnesses a great crowding of holdings into the "marginal" category. Percentage of holdings in this category has swelled from 49.8 per cent in 1970-71 to 76.7 per cent in 2002-03. According to National Sample Survey Report No. 491, percentage of marginal and small operational holding increased from 76.6 per cent in 1970-71 to 92.6 per cent in 2002-03 which covered only 64.9 per cent of total operated area while medium and large category of holdings constituted 6.9 per cent of land holdings and 15.3 per cent of total operated area. Average farm size has declined from 1.16 hectares in 1970-71 to 0.83 hectare in 2000-01.

In general, a rise is observed in the percentage of marginal holdings and a decline in percentage of holdings in all other categories. The pressure of growing population on the limited land base and the consequent division of holdings is clearly reflected in the changes in the absolute numbers of operational holdings in different size classes. It may be noted that numbers of operational holdings in different categories are not changing at the same rate, or even in the same direction, over time. Over the three decades the number of marginal holdings has increased from 10.4 million in 1960-61 to over 16.6 million in 2000-01, an increase of over one and half times. As against this, the absolute numbers of large and medium holdings have declined steadily during this period from 0.84 million to 0.49 million. Even the number of Semi-medium holdings, which had remained stable around at 1.5 million from 1970-71 to 2001-02, , has begun to fall.

Objective, Data Source and Hypothesis

The main objective of this chapter is to identify the size of an economic holding in Uttar Pradesh context. Size of an Economic Holding has been identified with the help of Cost of Cultivation data, Monthly Per Capita Consumer Expenditure, Household Consumer Expenditure among Socio Economic Groups; Yield Per Acre and Minimum Support Prices of Paddy, Wheat and Sugarcane Crops. This chapter also seeks to investigate whether the process of marginalization of holding has led to a decline in productivity.

The basic hypothesis of our study is that size of an economic holding varies with crops to crops and region to region in Uttar Pradesh. Our basic contention is that the small and marginal holdings suffer from a basic resource constraint, are unable to subsistence and cannot invest in modem inputs to the optimum extent. Though these holdings have an advantage over large holdings in terms of availability of free family labour and qualitative managerial and supervisory capacities, these cannot fully substitute the requirement of modern inputs. The access of institutional credit and support from government agencies in terms of input supply, extension, etc., to small farmers is also relatively limited. In other words, we hypothesize that the existence of an Economic Holding is larger than average size of holding as a whole.

Data and Methodology

The study is primarily based on analysis of both secondary and primary data. The data set used includes land distribution over different farm size categories and selected indicators of agricultural development at State level. The main data sources include Cost of Cultivation data, Ministry of Agriculture; NSS Reports on Household Consumer Expenditure among Socio Economic Groups; Yield Per Acre of Farm and Minimum Support Prices of Paddy, Wheat and Sugarcane Crops, Reports of the Commission for Agricultural Cost and Prices for 2003-04. Most of the data set related to 1970-71 to 2003-04.

The methodology used is based on simple mathematical analysis with the Cost of Cultivation Data and Monthly Per Capita Consumer Expenditure (MPCE) 2003-04 at the State level and across different size of farmer's category. Similar practice had been carried out by Khusro (1973). We first begin with a brief discussion of the pattern of landholdings in U.P., our study area.

Landholding Pattern in Uttar Pradesh

Agriculture is a State subject and thus policy for land reforms has differed across the country, though it has been guided by common national objectives. The success and impact of land reform measures has also varied across the States. Uttar Pradesh along with rest of the States in the country initiated the land reform measures soon after Independence. The measures included abolition of intermediaries, consolidation of fragmented holdings and redistribution of land through land ceiling laws.

The impact of these reforms on rural economy of Uttar Pradesh has been mixed. Abolition of *zamindari* provided an incentive for the peasant to invest in agriculture. The exploitative intermediaries were removed and the cultivator was brought in direct contact with the government. A necessary prerequisite for agricultural development was thus provided. Consolidation of land holdings has been by far the most successful land reform measure in Uttar Pradesh. It greatly contributed to the high growth rate in the 1970's. Land ceiling and redistribution of land, however, failed to provide the desired results, that of an equitable distribution of land. Only a small proportion of acquired land was redistributed and that too was of poor quality and paltry size (Singh, 1992). The average size of holding in Uttar Pradesh has been reduced to a highly uneconomical size of below one hectare. As per the NSS data 2002-03 three-fourths of the holdings in U.P. are below 1 hectare. Holdings below one hectare now account for almost one-third of the holding area. Around one-fourth area is under holdings between 1.0 and 2.0 hectares. Only one-thirteenth of total holdings in the State are above two hectares, though they account for about one-third of the total operated area.

Table 9.1: Changes in the Size Distribution of Operational Landholdings/ Operated Area/Average Size of Landholding (Uttar Pradesh, Rural)

Category of Land-holdings (in hectare)	*70-71 (in %)*			*81-82 (in %)*		*91-92 (in %)*		*2002-03 (in %)*		*2002-03*
	OH	*OA*	*AHS*	*OH*	*OA*	*OH*	*OA*	*OH*	*OA*	*AHS*
Marginal	49.8	15.6	0.36	59.6	18.1	68.0	25.0	76.7	35.7	0.38
Small	26.8	25.3	1.40	21.6	23.8	18.5	26.2	15.9	29.2	1.38
Semi-Medium	16.5	29.8	2.74	12.9	28.0	9.9	26.3	5.6	19.8	2.69
Medium	6.2	23.3	5.74	5.4	23.6	3.3	18.2	1.7	12.5	5.43
Large	0.7	6	16.11	0.5	6.5	0.3	4.3	0.1	2.8	13.94
Gini's Coefficient	0.471		1.16	0.520		0.498		0.450		0.80

Note: Marginal (0.02—1 hectare), Small (1–2 hectares), Semi-medium (2–4 hectares), Medium (4–10 hectares).

OH = Operational Holding, OA = Operated Area, AHS = Average Holding Size

Source: GOI, NSS Report No. 492: Some Aspects of Operational Land Holdings in India, 2002-03: 24

Table 9.2. Trend of Landholding ('000) Uttar Pradesh

Category of Holding	*1970-71*	*1980-81*	*1985-86*	*1990-91*	*1995-96*	*2000-01*	*CAGR (1971-2001)*
Marginal	10453	12582	13702	4819	16237	16659	1.55
Small	2689	2898	2964	3118	3136	3087	0.46
Semi Medium	1652	1614	1582	1543	1585	1427	–0.48
Medium	733	661	602	549	532	463	–1.50
Large Holding	112	72	55	45	39	32	–4.05
Total	15639	17827	18905	20074	21529	21668	1.08

Source: http://planning.up.nic.in, Annual Plan 2009-10 Vol. I, Part II, Date 27.7.2011.

Note: CAGR (Compound Annual Growth Rate)

Formula: Compound Annual Growth Rate (CAGR) computed as follows:

Compound Annual Growth Rate (CAGR) .

$$= \left\{ \left(\frac{\text{Ending Value}}{\text{Principle Value}} \right)^{\frac{1}{\text{time in year}}} - 1 \times 100 \right\}$$

Table 9.1 shows the values of Gini's coefficient which indicate land concentration has increased upto 1991 afterwards decreased to some extent in Uttar Pradesh. In other words land concentration is the same as earlier. Average operational landholding has reduced over time owing to high population growth rate, inheritance law, new concept of family system, abolition of village industries, rural indebtedness, social status with land ownership, social and economic forces, the system of crop sharing, unfinished land reforms programme like *zamindari* abolition, land ceiling, land consolidation and tenancy reforms. It is important to understand that compound annual growth rate (CAGR) of holdings is half of the CAGR of population during 1971-2001 in U.P which is different from all India scenario of marginalization of holdings There is great concern of economic viability regarding tiny holding because average size of landholding decreased gradually; although economic viability may vary with crops to crops, soil to soil and technology to technology. In Uttar Pradesh 92.6 per cent agricultural operational landholdings could not provide subsistence to the marginal or small farmers owing to tiny holding. In 1960s and 1970s; there emerged a question about plough unit, work unit, and income unit regarding marginal and small farm size (Khusro, 1973: 50).

This chapter is in five sections. In Section II, we explore the criteria of economic holding. In Section III, the size of economic holding is set up through secondary data. In Section IV, factors that make small farms viable are explored. In Section V, suggestions are given to make small farms economically viable.

Economic Holding

Economic holding is an expression that is much is vogue in agricultural economies but no precise definition has been attached to it. Smallness of the size of holding is generally associated with uneconomic holding and *vice versa.* But, in fact, small holdings are not inherently uneconomic nor a large holding is necessarily economic. The 'economy' of a holding is dependent on factors influencing the productivity of land and prices of farm output *vis-à-vis* costs of inputs and prices of consumption goods entering the farmer's budget. With an increase in productivity and in the prices of farm outputs relative to prices of consumer goods, a small holding can meet the tests of economic holding. In other words, the concept of economic holding is a dynamic one (Mehra, 1969:1523-1529). It is great concern that average farm size decreases over the time as a size below which results in the benefits or output obtainable from cultivation would be worthless. The recognition of the existence of uneconomic holdings may induce corrective steps that are feasible within the general framework of the economy. It would also aid in framing legislation to stop the continual reduction of the size of agricultural holdings through the process of inheritance.

According to Keating "Economic holding, which allow a man the chance of producing sufficient to support himself and his family in reasonable comforts after paying his necessary experience" (Joshi, 1992: 230). Size of the economic holding is quite a difficult phenomenon because it may varies from region to region depending upon nature of soil, level of rainfall, irrigation facilities, types of crops, price of the inputs and other climatic factors. In a region, where assured water supply is available to a fertile, soil, the size of economic holding will be small as compared to the same in dry and semi-arid region. U.P. Congress Committee (1950) has defined economic holding like, basic and optimum holding. An economic holding is that which offers a reasonable living standard to the family of a farmer, provides full employment to family members with a pair of bullocks and it would be in relation to factors of agrarian economy of region. Khusro (1973) suggested some economic holding measures like Plough Unit, Work Unit and Income Unit regarding some States like Uttar Pradesh, Andhra Pradesh, Punjab, and Madras.

Some Concepts of Economic Holding

Not all farms and farmers in India are suffering from diseconomies of scale; neither small nor all farms and farmers subject to tenurial disincentives. Real concern here is precisely what proportion of farm families and farm acre is subject to size disabilities and tenurial disincentives and what proportion is free from these handicaps. Such a demarcation might isolate the areas and farm families which have necessarily to wait upon a programme like diversifications, water revolution and second green revolution. The procedure followed here is to arrive first at some estimate of farm size below which a farm is palpably inefficient and uneconomical. In doing somewhat detailed examination of the concept of plough unit, work unit and income unit concepts are applied to the farm management data of various States.

Plough Unit—Below which a single pair of bullock remains underutilized. **Work Unit**—A size of farm below which family labour will be underutilized. **Income unit**—A size of farm must be fully absorbed to the family manpower including managerial labours in gainful employment even if we allowed for the seasonal nature of agriculture (Khusro, 1973, p. 38). **Basic Holding**—A size of farm which is smaller than the economic holding and that could provide subsistence to the farmer. However, the basic holding is not sufficient for the purpose of agricultural production on large scale and use of modern technology. **Optimum Holding**—A size of holding which is three times of economic holding as per Agrarian Reforms Committee (Joshi, 1992, p. 230). **Family Holding**—A size of family holding is an area equivalent, according to local situation and under existing technical knowledge that provides certain income level. These measures have been adopted by Government of India during Plan period for land reforms in Indian agriculture. But these measures and definitions cannot be applied in contemporary agricultural condition.

Table 9.3: Plough Unit, Work Unit and Income Unit (in acre)

State	*Plough Unit*	*Work Unit*	*Income Unit*
Andhra Pradesh	10	5	10
Madras	5	7.5	15
Punjab	5-7.5	10	20
Uttar Pradesh	10	7.5	10
All India	7.5	7.5	15

Source: Khusro (1973, p. 50, 60, 67).

Table 9.3 shows that approximately 7.5-10 acres of land represent Plough Unit and Work Unit. Whether 10 acres represents the Income Unit? This is the dividing line between inadequate and adequate family employment. These findings are based on farm management data (1954-55) in the case of selected States. In hectare, plough unit and work unit requires 3-4 hectares and Income Unit requires 4-6 hectares (Khusro, 1973, p. 50, 60, 67).Table 9.1 also reveals that 2.7 hectares need to maintain Plough Unit and Work Unit and 4.0 hectares need to maintain Income Unit of operational landholding. If we follow Khusro's finding, 98.2 per cent operational landholding become unviable in Uttar Pradesh in terms of Plough Unit and Work Unit. Only 1.8 per cent holding accounting to 15.3 per cent operated area only represent Income Unit in this case in present circumstances as average land size holding of marginal farmers in the State is 0.40 hectares, consequently this comes to 1.41

hectares for small, 2.74 for semi-medium, 5.57 for medium and 15.07 for large.

Economic Holding Definition

It is very important to note that plough unit and work unit is fly-by-night concept owing to existing new technology like tractor, tube-well, and thresher machines. Based on above review of literature we may define an economic holding as a unit which should provide subsistence to all family members at current prices in present situation and also fulfil the cost of cultivation excluding input cost and net Income from cultivation should fulfil subsistence of a family households.

Bearing in mind the difficulties in the estimation of costs of cultivation and the income of the household accruing from both farm activities and non-farm activities, an economic holding may be alternatively defined as the size of land per household which, besides meeting material cultivation costs, consumption expenditure for the family workers including family members of the household, without necessitating, on their part, extra labour outside cultivation for meeting their own expenses for livelihood. In short, economic holding represents a size of land per household which should at least meet both consumer expenditure of family members and the cost of material inputs, seed, feed, manures, irrigation charges, agricultural impediments, pesticides, etc., incurred by the households. In other words, we may define an economic holding through which cost of cultivation (A1) and consumption expenditure of a household is incurred, excluding family members who engage in off-farm activities.

Average total monthly income of the farmer household as per Table 9.4 was Rs. 1,633 (annual income Rs. 19,596). It shows that among the four sources considered, cultivation is the most important source of income (51.1%) per farmer household, from wages it was Rs. 559 while income generated from non-farm business was Rs. 185 and income from farming of animals was only Rs. 53. However, there is large difference in the total income across size classes. The total income from cultivation for the farmers up to 1 hectare was much lower than the other farm size classes. It means 98.1 per cent farmer households do not get subsistence from cultivation. So there is need to improve productivity on marginal farms through the diversifications and best farm management practices.

The question is whether the farm cultivation is sufficient to meet all the basic necessities of a family farm. One household needs more than Rs. 2175.7 per month in order to cross poverty line in U.P (GOI, Planning

Table 9.4: Average Monthly Income from Different Sources, Consumption Expenditure and Net Investment in Productive Assets in Rs. per Farmer Household in Uttar Pradesh during Agricultural Year July 2002 to June 2003

Size Class of Land Possessed (hectares)	*Income from Wages (Rs.)*	*Net Receipt from Cultivation (Rs.)*	*Net Receipt from Farming of Animals (Rs.)*	*Net Receipt from Nonfarm Business (Rs.)*	*Total Income from different Sources*	*Total Consumption Expenditure (Rs.)*	*Saving (Rs.)*	*Net Investment in Productive Assets (Rs.)*
< 0.01	873 (75.9)	1 (0.08)	–7	283	1150	2235	–1085	57
0.01-0.40	654 (56.9)	232 (20.2)	39	223	1148	2354	–1206	42
0.41-1.00	376 (29.4)	696 (54.4)	53	153	1278	2831	–1553	101
1.01-2.00	501 (20.63)	1746 (71.9)	63	118	2428	3728	–1300	211
2.01-4.00	596 (14.98)	2999 (75.3)	189	194	3978	4948	–970	631
4.01-10.00	545 (8.0)	7282 (91.3)	218	–71	7974	6732	1242	1252
>10.00	72 (0.9)	8370 (106.6)	–774	182	7850	6776	1074	228
All sizes	559 (34.23)	836 (51.1)	53	185	1633	2899	–1266	136

Note: Income as percentage out of total income in bracket.

Source: NSS Report No. 514, p. A-190: Household Consumer Expenditure among Socio-Economic Groups: 2004-05.

Commission, 2009: 17). It implies that 93.3 per cent farmer households do not get adequate Net Farm Income from cultivation to fulfil their family needs. Nearly 60 per cent of farmers do not access any information on modern technology from any source (Dev, 2008:37).

Cost of Cultivation and Economic Holding

The most important issue is the nature of cost and price of output and difference between total cost of inputs and total price of farm output. Instead of this, we should also measure that difference is rationale or not. It is also important that imputation of own account labour and owned investment cost. In fact, if we account wages of family workers at the prevailing wage rates of agricultural labourers, the profitability of farm enterprise in many holdings is found to be negative. Since the farmer is a residual applicant in the proceeds of the farm enterprise and meets part of his deficit in expenditure through wage employment within and outside agriculture, the estimated returns would be still less.

Therefore, though the 'economy' of size of holding should not be rigidly identified with any particular size of land, there is yet needed to have a working definition of economic holding. Various concepts of economic holding may be attempted. To start with an economic holding may be taken to as size of land per household, the gross income from which at least the total cost of cultivation is covered. We may use the cost of cultivation in the form of Cost A1, Cost A2, Cost B1, Cost B2, Cost C1, Cost C2, and Cost C3 for various crops. That is, the income obtainable from the sale of the produce of land operated by the household covers at least the cost of all material inputs, management costs and service charges. On the theoretical plane this definition appears quite sound. But certain practical issues emanate from it which renders the task of the estimation of the size of the economic holding quite difficult. The quantification of the size of economic holding is a complicated issue, if not in theory, surely in practice. In this chapter issues involved in the measurement of the size of economic holdings are discussed, and in problematic cases, the concept of the size of economic holding is suitably modified and workable substitute techniques are suggested.

Measurement of Economic Holding of Uttar Pradesh

If a farmer grows Paddy and Wheat on particular unit of land then the income obtainable from the sale of the produce of land operated by the farmer household covers at least the cost of all material inputs and Total Consumer Expenditure of a Farmer household in Year or Net Income from cultivation should be equal to Total Consumer Expenditure in a Year.

We have used the data of Cost of Cultivation (A1)[4] for one hectare and Minimum Support Prices (MSP) in Rs. / quintal, Monthly Per Capita Consumer Expenditure (MPCE) in U.P. for the year 2003-04.

Assumptions

1. Farmers are cultivating Paddy and Wheat or Sugarcane only in a year.
2. Marginal and Small farm is family labour intensive.
3. Most of cultivated area is irrigated.

We may write a function of economic holding as:

$$E = f(cc, ce, y, df)$$

Where, E = Economic Holding, cc = Material cost of inputs, ce = Consumption expenditure of a farmer household, y = Yield, df=Dependent family members of a farmer's households

Relations between variables: $E \propto 1/y$, $E \propto cc$, $E \propto ce$, $E \propto df$

- Size of an Economic Holding = TCEY/ NFIpw
  (Where, TCEY= Total Consumer Expenditure of a Family in a Year = MPCE × Household Size × 12; NFI= Net Farm Income = Total Revenue – Cost of Cultivation (COC)

Total Consumption Expenditure of a Farmer Household in a Year (TCEY[3]) in U. P. = 532.63 × 6.5 × 12 = 41545.14 ...(i)

Net Farm Income = Total Revenue from Cultivation of Paddy and Wheat – Cost of Cultivation (COC) of Paddy (p) and Wheat (w)

TRpw = TRp + TRw

TRpw = MSPp × Output per Hectare of Paddy + MSPw × Output per hectare of Wheat

TRpw = 550 × 38.07 + 620 × 34.01 = 42024.7 ...(ii)

COCpw = COCp + COCw

COCpw = 8696.98 + 10407.38 = 19104.36 ...(iii)

(*Note*: Cost A1—All actual expenses in cash and kind incurred in production by owner)

NFIpw = TRpw – (COCpw) or NIpw = (ii) – (iii)

NFIpw = 42024.7 – 19104.36 = 22920.34

Size of an Economic Holding = TCEY/ NFIpw

= 41545.14/ 22920.34 = 1.81 hectares

Size of an Economic Holding regarding Paddy and Wheat in U.P. = 1.81

If a farmer household cultivated only Sugarcane:

Cost of Cultivation (COC) of Sugarcane = 10547.01 ... (iv)

TRs = MSPs × Yield/hectare

TRs = 73.00 × 490.36 = 35796.28 ...(v)

NFIs = TR – COCs = (v) – (iv)

NFIs = 35796.28 – 10547.01 = 25249.27

Size of an Economic Holding = TCEY/ NFIs

= 41545.14/25249.27 = 1.64 hectares

Size of an Economic Holding regarding Sugarcane in U.P. = 1.64 hectares

We found that 2 hectares or 5 acres may represent an economic holding in U.P. On the other hand the adoption of HYV seed, irrigation and access of small holders to power from government extension facility, increased per acre output over time, leads to satisfaction of that norm with a smaller acreage. Considering that U.P is one of the State with low per capita income and low productivity, it is no means fantastic to think that a threefold to fourfold productivity would increase in a short span of time. In fact in U.P, productivity per acre more than doubled within two decades, and will probably doubled again in the following decade along with millions of others farms. Costs of input per hectare will, of course, increase but even so net productivity per acre could rise by a very significant magnitude especially with the symptoms of an agricultural transformation emerging in many places. As that happened, the unit of land which gives adequate family income will shrink below its present magnitude which, as we saw, is stated, perhaps with exaggeration, to be 2 hectares on the average. We may then expect the income unit of tomorrow to be moving in the direction of 1.5 hectares or perhaps even 1 hectares which is near average farm size of U.P.

Based on above equation it is clear that there is inverse relationship between size of an economic holding and yield and direct relationship with the cost of cultivation, Consumption expenditure, and dependent-family members on farm activities. Therefore, findings suggest that if we increase output per unit of land, size of an economic holding decreases over the time period and decreases cost of cultivation per unit of land or decreases number of dependent family members or decreases size of a farmer household, size of an economic holding decreased and farms become viable. We may say that Economic Holding in Uttar Pradesh may average 2 hectares per household. It may vary with cropss to crops or region to region in the State. According to Jagdeesh Gangwar, a small farmer of Bareli district, Uttar Pradesh: "first we grow paddy and wheat as per needs of family members then we cultivate cash crop like sugarcane on rest of the land and if we have 2 hectares or 5 acres land, we easily survive".

It is important that size of agricultural land holding is not much important; our concern is Farm-Size productivity which is in favour of small size of farm. It is well known that there is inverse relationship between farm size and output per unit of land. However, the recognition of the factors which increase productivity, the new agricultural technology with traditional practices (using HYV seed, fertilizer, manure, bio-fertilizer etc.) and their acceptance by the farmers involve a lagged relationship and, insofar it is so, the static concept is equally meaningful. There is, therefore, a need to have a concept of economic holding which serves as base for measuring changes in the economic characteristics of different holdings as also for correcting errors emanating from the adoption of unreal standards in determining economic holdings. The recognition of the existence of uneconomic holdings may induce corrective steps that are feasible within the general framework of the economy. It would also aid in framing legislation to stop the growing minifundias through the transfer of land to inheritance (the word minifundia is used for very small farms in international encyclopedia of Economics, Vol. I, 1997: 840).

At the national level value of output per hectare of Marginal, Small, Semi-Medium, Medium Farmers are Rs. 14754, Rs. 13001, Rs. 13944, Rs. 11333, and Rs. 12535 respectively (NCEUS Report, 2008). There persists inverse relationship between farm-size and productivity in Indian agriculture and Uttar Pradesh (Sen. 1962, 1964; Khusro, 1964; Rao, 1966; Saini, 1971; Patnaik, 1972; Bhardwaj, 1974, Chattopadhyay and Sengupta, 1999; Dogra, 2002; Chand, Luxmi and Aruna, 2011, Gaurav and Mishra, 2011) and my own field work also reckoned inverse relationship between farm size and productivity. The supporters of such inverse relationship between farm-size productivity have given mainly two types of argument: (i) intensity based and (ii) quality based.

One of the arguments to explain inverse relationship between farm size and productivity is that the family labour is cheaper than the hired labour. Moreover, family labourer works hard, carefully and makes economic and proper use of other inputs like seeds, fertilizers, equipments etc. It is family labour which is engaged more on small sized farms. Hired labourers enter as a specific cost of production. Another side argument is quality based, which implies that small farmer's family depends on a small size of land and only their best effort can survive them. These situations force them to improve the quality of their land and try to cultivate more than one crop from the small-piece of land. This attitude increases per acre production high.

From the above discussion we can understand the importance of small size farm regarding productivity in Uttar Pradesh that there is need to make a small farm viable.

Remedies for Economic Holding

Different variety of remedies have been considered and applied to solve the problem of small sized farm in the Uttar Pradesh. These remedies can be grouped into two heads (i) Preventive measures and (ii) Curative measures. The former are about such measures, which stop further division of farms. While, curative remedial measures are applied to reduce or to abolish evil of small farms. Several measures have been adopted by the State to prevent subdivision of farms. It is necessary to develop such system in which land is nominally divided among heirs, but left in the hands of one person to cultivate for all the members. There must be prohibition of fragmentation of land below a certain limit. To avoid further fragmentation of land holding, joint farming system or circle order farming system without partition can be developed.

There is urgent need to restart curative measures to solve the problem of small-sized holdings. One of them is consolidation of holdings. This measure may increase output per unit of land and acquire economies of scale. In the present situation, farmers of Uttar Pradesh are agreeing to consolidate their agricultural land (observation from field survey). Eight Factors emerged to make a small farm viable with sustainability: to increase access capacity of small farmers and revive working system of all government agricultural institutions, to set up a Soil Testing Centre in every *Panchayat* and provide suggestions and remedies according to soil quality, Start Canalization, Eliminate Distress Sale, Protect Animal Husbandry to get manures, Credit direct to Small Farmers, protect crops from being affected by worms and to save pulses crops from blue bull, to provide timely fertilizer and electricity to the small farmers.

Rain water has great importance in terms of irrigation through groundwater resources, the main source of irrigation, is in jeopardy on account of the falling groundwater level. If assured irrigation is provided, labour absorption within the agricultural sector will also increase as is the case in many countries. There is 71.6 per cent area coverage under the irrigation in Uttar Pradesh (*Agricultural Statistics at a Glance*, 2008: 93). A *Pradhan (Mukhia),* in a village of Raebareli District, Uttar Pradesh said that if government provides timely irrigation, fertilizer and power supply then village will double the output per unit of land[1].

Distress Sale: Most of farmers (Marginal and Small) in U. P have been selling their product to the middleman supposing at the rate of Rs. 1000 per quintal and middleman sold it to government selling and purchase center at Rs. 1170 per quintal and they also give commission at the center Rs. 40 per quintal to the Center's chairman. It is well known that in the absence of rural godowns, marginal and small farmers are forced to sell to the middleman at very low prices as they have no holding capacity. The cooperative rural go-downs can arrange finance on the collateral of the stored produce to meet the immediate needs of the farmers which will obviate the necessity to sell to private traders and middlemen. It is important to know that middlemen and private traders has been providing bridge loan to the marginal and small farmers against future crop. If cooperative/government agencies/commercial banks provide bridge loan as collateral; distressed sale will slow down[1].

Animal Husbandry: The development of it provides great possibility in the State. In absence of animal husbandry farmers are not able to use adequate dung/manure in their field which contribute to the fertility of the soil by adding organic matter and nutrients, such as nitrogen, that are trapped by bacteria in the soil. Higher organisms then feed on the fungi and bacteria in a chain of life that comprises the soil food web. This is a known fact that mixture of dung and fertilizer make soil more productive. The resource poor can start some remunerative activity with small capital but the situation is such that there is deep poverty, people may not afford to keep even a goat.

Conclusion and Way Forward

Uttar Pradesh agriculture sustains a large cultivating population on the meager land resource. Land ownership distribution remains highly skewed and average size of landholding has been steadily declining. An overwhelming majority of farmers are marginal and small cultivators. These farmers use available inputs intensively, except in some regions, show higher efficiency. In general, small farms are not less efficient than large farms. However, due to a low resource base and inadequate income from alternative sources, small farmers are caught in a vicious circle of poverty and have to be supported in their relentless struggle to attain their full potential and have a better future.

We may explain 'small farmers: decline and persistence' in this way that is Poverty of Small Farmer-Households: Small quantities of Land and Capital? Low Capacity to Invest in Farming → Small Total Production→ Weak Bargaining Power → Poor Access to Public Extension Services → Inadequate Access to Inputs, Markets and Credit, Minimum Support Prices, Due to Lax Administration, and inadequate infrastructure of Non-Farm Employment → Low Income → Result Poor Small and Marginal Farmers.

An attempt is made in this chapter to understand that there are two ways to provide subsistence to the small farmers and make viable farms in present agrarian situation in Uttar Pradesh. The first argument is to increase land productivity or decrease cost of cultivation through remedial measures as suggested above and increased awareness regarding government's programmes including existing machinery and initiation of new farming techniques in favour of small and marginal farmers. Second argument is to evolve appropriate non-farm activities in rural areas which may be centred on agricultural inputs and small scale industry.

This implies that strategies/ policies may be targeted to increase household income in rural areas by increasing access to land resources coupled with promoting off-farm activities. Such diversity to an important extent may be determined by economic mechanism which is influenced by initial condition and reform policies, focus on non-traditional farm structure during transition as it is optimal to address the specific institutional and structural constraints imposed by the transition process.

The government may revive a targeted programme focusing on small and marginal farmers, with an initial thrust on areas where the existing yield gap is considered to be high. A special agency or coordinating mechanism may be set up for this purpose in every State as per need in Twelfth Five Year Plan.

We found that 93.3 per cent of farmer households have not been getting adequate income from farm activities for subsistence. So, there is need to provide basic infrastructure related to non-farm/small scale industry in villages which would increase income of small farmers and agricultural labourers. The agrarian structure is in flux. There is inexorable downward pressure on farm size and increasing concentration in the lowest size group and increasing inequality in agrarian structure. Increasing migration and occupational restructure are consequent responses discernible in the dry region side particularly in Bundelkhand and East U.P. also.

Expansion of farm incomes is still the most important weapon for reducing poverty. Non-farm income opportunities notably, horticulture, livestock, dairying, poultry, fisheries post-harvest operations, maintenance of farm equipment, etc., offer a virtuous cycle connecting expansion of farm activity to that of rural non-farm income opportunities.

NOTES

The observation has been carried out from field work of my D.Phil. work during January-April, 2011.

Personal Interview of *Gram Pradhan* during D.Phil. field work.

TCEY is computed from NSS report No. 514: Household Consumer Expenditure among Socio Economic Groups, 2004-05

It is taken into consideration that size of a household in U.P is 6.5 (as per Census 2001: 10).

Amount of Cost and Cultivation for 2003-04 is taken from Report on 'Cost of Cultivation of Principal Crops in India' Ministry of Agriculture, GOI (2007).

Concept of Cost:

Cost	*Items*
Cost A1	All actual expenses in cash and kind incurred in production by owner.
Cost A2	Cost A1 + rent paid for leased in land.
Cost A2 + FL	Cost A2 + imputed value of Family Labour.
Cost B1	Cost A1 + interest on value of owned capital assets (excluding land).
Cost B2	Cost B1 + rental value of owned land (net of land revenue) and rent paid for leased in land.
Cost C1	Cost B1 + imputed value of Family Labour.
Cost C2	Cost B2 + imputed value of Family Labour.
Cost C2*	Cost C2 estimated by taking into account statutory minimum or actual wage whichever is higher.
Cost C3	Cost C2* + 10 % of Cost C2* on account of managerial functions performed by farmer.

Source: GoI (2011), Reports of the Commission for Agricultural Cost and Prices (2010-11), Ministry of Agriculture, p. 369.

Acknowledgement: The author is pursuing D. Phil. work on "Size-Differential in Agricultural Land Holding and Its Impact on Productivity: A Study Based on Selected Farms in Uttar Pradesh" under the supervision of Prof. Bhaskar Majumder from Govind Ballabh Pant Social Science Institute, Allahabad, Uttar Pradesh. The author is thankful to my Ph. D guide Prof. Bhaskar Majumder, Prof. M. Muzzamil and Prof. Ravi Srivastava who gave me valuable suggestions in writing this paper.

REFERENCES

Chand, Prasanna and Singh (2011). "Farm-Size and Productivity: Understanding the Strengths of Small Holders and Improving their Livelihoods", *Economic and Political Weekly,* Vol. XLVI (26 and 27): 5-11.

Chattopadhyay, Manbendu and Atanu Gupta (1999). "Farm Size and Productivity" *Economic and Political Weekly*, Vol. XXXIV (19): 147-1148.

Dev, Mahendra S. (2008) "*Inclusive Growth In India: Agriculture, Poverty, and Human Development",* New Delhi: Oxford University Press.

Dogra, Bharat (2002) "Land Reforms, Productivity and Farm Size", *Economic and Political Weekly,* Vol. XXXVII (6): 532-533.

Gaurav, Sarthk and Srijit Mishra (2011). "Size- class and Returns to Cultivation in India: A Cold Case Reopened", Working Paper, IGIDR, Mumbai.

GoI (2008) "*Agricultural Statistics at a Glance*", New Delhi: Directorate of Economics and Statistics, Ministry of Agriculture.

GoI (2008) "Report on a Special Programme for Marginal and Small Farmers", National Commission for Enterprises in the Unorganized Sector: New Delhi.

GoI (2011) "*Draft Faster Sustainable and More Inclusive Growth: An Approach to the Twelfth Five Year Plan (2012-17)"* New Delhi: Planning Commission.

GoI (2011). "Reports of the Commission for Agricultural Cost and Prices: 2010-2011", CACP, Ministry of Agriculture, New Delhi, p. A-143.

http://www.mospi.gov.in, *NSS Report* No. 407, 491, 492, and 515.

Joshi, B. H. (1992). *Problems of Indian Agriculture: A State-wise Study*, B.R. Publication, Delhi.

Khusro, A. M. (1973). "*The Economics of Land Reform and Farm Size in India*", New Delhi: Macmillan.

Mehra, D. S (1969). "Economic Holding: Concept and Measurement" *Economic and Political Weekly*, 4 (38): 1523-29.

NSS Report (2003). "Income, Expenditure and Productive Assets of Farmer Households", Report No. 497, p. A-190.

NSS Report (2004-05). "Household Consumer Expenditure among Socio Economic Groups":, Report No. 514, p. A-143.

Patnaik, Utsa (1972). "Economics of Farm Size and Farm Scale", *Economic and Political Weekly*, Vol. VII (31 and 33):1613-1624.

Sen, A. K (1962). "An Aspect of Indian Agriculture", *Economic and Political Weekly,* 14 (4, 5, and 6): 243-247.

Sen, Amartya (1964). "Size of Holding and Productivity", *Economic and Political Weekly, February: 323-326.*

Shankar, Kripa (2011). "Central Budget 2011-12 on the Context of the Rural Sector", *Mainstream,* 44 (14). March 26, 2011.

Singh, Ajit Kumar (2005). "Role of Rural Non-Farm Sector in Rural Transformation: Evidence from Uttar Pradesh" in Rohini Nayyar and Alakh N. Sharma (Eds) *Rural Transformation in India: Role of Non-Farm Sector*, New Delhi: Institute for Human Development.

Singh, Ajit Kumar (2008). "Land Reforms in Uttar Pradesh: Retrospect and Prospects", *IASSI Quarterly,* Vol. 10, No. 3, 1992.

10

Agriculture in Odisha: Prospects for Growth and the Policy Imperatives

Prashanta Ch. Panda and *Sudhakar Panda*

Why Develop Agriculture?

Odisha's economy has been changing and changing fast. The transformation of the economy from a subsistence agrarian economy to a modern economy is visible in terms of the composition of State Gross Domestic Product (SGDP) where the relative contribution of agriculture and allied activities has been declining and that of trade, commerce, industry and services is increasing. In 2009-10, agriculture contributed 18 per cent to the SGDP *vis-à-vis* the contribution of industry and the services sector stood at 28 per cent and 54 per cent respectively. In the coming decade contribution of agriculture to SGDP may decline further and that of other sectors including industry may increase. Yet the major defining trend of Odisha's economy, notwithstanding the advent of modernization and commercialization of the economy may remain agrarian for a long time for the following reasons:

The first is the assurance of food security which comes from the development of agriculture: the production of cereals like rice, wheat, bajra and millets constituting the major food crops for the people of Odisha. Added to it, relatively higher weightage of foodgrains in consumer's price index has restricted State's options to allow agriculture to move in the market oriented path of growth. Further, the role of agriculture in ensuring equitable access to food has forced food prices to be maintained at lower levels. This, in turn, has not created favourable conditions for the development of the farmers in terms of a diversified growth of the farm sector.

The second is the sheer scale of employment that agriculture offers today. As against 56 per cent of the workforce engaged in agriculture at the all India level, the share of the workforce engaged in agriculture is around 70 per cent in the State coupled with the fact that the pace of shift of population from agriculture to non-agricultural activities has been rather slow. The third reason is the importance of agriculture as the source of livelihood of the majority of the people living in rural areas. In fact, the bulk of Odisha's population around 80 per cent lives in rural areas and given the slow pace of change from agriculture to non-agricultural activities in the State, agriculture and allied activities will continue to be the centre of economic activities. This view is further strengthened by the fact that given the speed and shape of technological innovations and developments in the economy after globalization, chances of shifting of existing labour force from the rural sector to the manufacturing and services sector is remote unless they are empowered with requisite education, knowledge and skills, which is a long run focused process. It is only a developed and diversified agriculture sector propelled by technology and consumers' choices that can be expected to change the structure of production and the quality and size of employment as well. Changing tastes and preferences of the consumers with higher disposable incomes will shape the production strategy of the farmers.

Predominance of Small and Marginal Farmers

Our arguments for the development of agriculture become all the more compelling when we look at the structure of farm households in the State. It may be observed that as high as 85 per cent of the farmers in the State are small and marginal. If development of agriculture has to be inclusive and job oriented, priority needs to be accorded to these groups of farm households. Improvements in productivity and in the growth rates should benefit them as in no case the development of agriculture can be decoupled from their growth. Further, the farm studies conducted earlier in India indicated the higher production efficiency of the small farms *vis-à-vis* the large farms. This should strengthen our hope for the development of Odishan agriculture dominated by small and marginal farms.

Is Agriculture the Gateway to Odisha's Prosperity?

Agriculture in Odisha, given the present shape of socio-economic development in the State will continue as the most important sector of the Odishan economy. Agriculture not only creates employment and generates income, but also ensures self-sufficiency in food production. Rural population in the State is dependent upon agriculture even in cases where it is not remunerative and fraught with adverse scenario of under-employment. They have no other choice in the absence of alternative sources of employment even when agriculture in the State suffers from recurring floods, droughts, pests and pestilence, untimely rains and cyclone which result in ruining their prospects of income and livelihood. Added to it, water and soil management practices and crop management are undertaken in less scientific manner. Storage, processing and marketing activities are grossly inadequate.

Lack of detailed data in relation to a farmer, his land and cropping pattern and yield for *Kharif* and *Rabi* seasons for a number of years put a stumbling block in evolving an area approach for rational formulation of an insurance policy or even for a government support plan at the *Panchayat* or Block level. But the institutional efforts leading to the growth and corresponding changes in terms of infrastructure, research and development, credit support, and marketing are no where emerging in a scale to break the vicious circle that still prevails in State's agriculture. Yet, there is no escape from the development of agriculture as that lays the foundation for the growth of Odisha's economy and brings hope and prosperity for its people. The problem before us: how best we can do it?

What Matters for Agriculture Development

Access to institutional credit to farm households remains crucial for agriculture development. Availability of agricultural credit in appropriate quantity and quality (on the basis of purpose and time) is also equally important for realizing the true potential of agricultural engagements as remunerative activities. Farmers need credit for short period (up to 15 months) for the purchase of seeds, fertilizers, and fodder for livestock, etc., medium term (15 months to 5 years) for the purchase of agricultural tools and implements, cattle, repairing wells and also long term (more than 5 years) for the purchase of heavy machineries like tractors and harvesters. Money lenders are easily approachable by the credit seekers and they don't have to satisfy many formalities in transacting a loan. The money lenders may resort to malpractices pertaining to charging of high rate of interest. They may even adopt unfair means in maintaining the records of accounts. Though introduction of *Kisan* Credit Cards (KCC) for short term credit is helpful for the farmers, lack of updated land records, small and marginal landholdings and illiteracy of borrowers has limited the outreach of KCC. Self Help Groups and Bank Linkage Programme and other Micro Finance Institutional arrangements have not come up in significant manner to reach the target for small and marginal farmers, agricultural and non-agricultural labourers, artisans, craftsmen and other poor engaged in small business like vending and hawking. These with credit co-operative societies form an integral part of rural credit system.

More than 80 per cent of agricultural families (90 per cent of villages) have been enrolled as members of co-operatives. But less than 35 per cent of enrolled members are the servicing members having regular transactions with the credit co-operative. Market share of co-operatives in deposit mobilization is only 7 per cent as against 93 per cent by commercial and regional rural banks in Orissa. Cooperatives provide 64 per cent of agricultural credit under short term seasonal agricultural operations. Taking all credit portfolios for agricultural support, stake of cooperatives is between 50 per cent to 54 per cent. In recent years the importance of co-operatives in the supply of credit to the agriculture has declined and that of RRBs has gone up. Table 10.1 gives us an idea about the gross cropped area, production of foodgrains *vis-à-vis* the availability of credit in the State in 2001-02:

It may be seen from Table 10.1 that the highest share of credit disbursed, around 52.74 per cent has gone to the coastal plains followed by the Central Tableland at 20.37 per cent, Northern Plateau at 15 per cent and Eastern Ghats at 11.88 per cent respectively. Given the importance of credit as a catalytic factor in the development of agriculture, there needs to be a well designed strategy of linking farm credit not only with the production of crops but also with the marketing of crops. This may prevent distress sale of paddy and other crops and create motivation for the farmers for continuation and expansion of their agricultural operations. It is good to find instances where PACS have been selected as State agency for procurement of paddy under decentralized programme by the Food and Civil Supply and Consumer Welfare Department to help farmer members get remunerative price. A more positive approach to agriculture by the credit institutions including the commercial banks in terms of acceptability of the tenants for sanction of loans and other facilities can be expected to add to the stability and progress of this sector

Table 10.1: Region-wise Sharing of Gross Cropped Area, Production of Foodgrains and Agricultural Credit in Orissa during 2001-02 (Per cent)

Regions	*Gross Cropped Area*	*Production of Foodgrains*	*Credit Disbursed*	*Districts*
Coastal Plains	35.15	38.95	52.74	Balasore, Bhadrak, Cuttack, Jagatsinghpur, Khurda, Jajpur, Puri, Kendrapara, Nayagarh, Ganjam
Central Tableland	24.37	26.29	20.37	Baragarh, Sambalpur, Deogarh, Jharsuguda, Bolangir, Sonepur, Dhenkanal, Anugul
Northern Plateau	14.84	14.26	15.00	Mayurbhanja, Keonjhar, Sundergarh
Eastern Ghats	25.64	20.50	11.88	Gajapati and others in KBK

Source: *Government of Orissa, Economic Survey, Proceedings of SLBC, Orissa.*

and their stakeholders. It, therefore, becomes necessary to adopt a fresh approach to the problem and liquidate the unviable primary co operatives where necessary in the interest of sound norms and financial discipline in the credit co-operative sector. Supervision and monitoring of the network need to be more effective and vigilant so that the system becomes more responsive and sensitive towards its responsibility. Modern management and financial practices may also be adopted by the cooperative credit institutions to provide quality service to the customers. This may enable the co-operatives to have an edge over other players.

What is Wrong with Credit Supply?

Farmers often complain about the non-availability of credit in time and inadequate quantum of credit offered to them by the banks.Table 10.2 below indicates the size of credit disbursed per hectare of gross cropped area in the State *vis-à-vis* the country.

It may be observed from the data that agricultural credit disbursed per hectare in the State continues to be significantly lower as against the size of agricultural credit disbursed per hectare at the national level. It

Table 10.2: Agricultural Credit Disbursed Per Hectare of Gross Cropped Area in Orissa *vis-à-vis* India

(*In Rupees*)

Year	Orissa	India
1994-95	278.61	1022.02
1995-96	390.98	1201.30
1996-97	536.76	1337.19
1997-98	541.35	1699.18
1998-99	718.35	2074.92
2000-01	1035.79	2784.76
2001-02	1055.92	3265.52
2002-03	1333.25	3622.91
2003-04	1536.41	3947.45

Source: Report on Currency and Finance, RBI, various issues, Economic Survey, Government of Orissa, Proceedings of SLBC, Orissa.

may also be noted that the gap in agricultural credit disbursement between the State and as observed at the national level continues in a bigger way right from 1994-95 onwards. This is as much a reflection on the agricultural credit policies of the banks as it is reflection on the credit absorbing capacity of the agricultural sector. This may also be taken as an indication not only of the low level of agricultural development in the State which constrains the flow of credit to this sector but also of the poor record of repayment of loans and interests by the borrowers. This creates a big handicap for the institutions to invest more funds in the agricultural sector and suffer great losses. An attempt is made in Table 10.3 to know the real situation relating to credit outstanding in different States including Odisha.

Figures depicted here make it clear that credit outstanding per hectare of net sown area since 1972 against all the States has been increasing steadily. In case of Odisha, it has been consistently rising from Rs. 72 in 1972 to Rs. 448.00 in 1982 to Rs. 1103.00 in 1995-06 and to Rs. 2039.00 in 2001-02.The situation is no different for other states.

As against the growing indebtedness of the farm households in the agricultural sector, one has to view the recurring incidence of pests, flood, cyclone, earthquake and hailstorm which cause extensive damage to crops. Premium received per farmer in crop insurance is the highest for Andhra Pradesh, almost twice that of Orissa (Rs. 440.62 *vis-à-vis* Rs. 221.46). There are twenty two types of crops covered in Andhra Pradesh as against eight in Orissa. The wide range of crops cultivated in Andhra Pradesh

Table 10.3: Credit Outstanding Per Hectare of Net Sown Area in Different States

(*In Rupees*)

State	*1972*	*1982*	*1995-96*	*2001-02*
Andhra Pradesh	197	815	3519	7146
Haryana	145	1168	2767	7831
Karnataka	124	605	2719	6500
Kerala	283	2486	5583	12204
Punjab	289	740	4073	9578
Tamilnadu	310	1369	6292	9191
Uttar Pradesh	120	593	1735	4127
West Bengal	56	543	1651	3413
Gujarat	237	554	1614	3873
Orissa	72	448	1103	2039
Madhya Pradesh	69	277	1163	2520
Bihar	59	383	1955	2009
Rajasthan	31	202	691	2639
All India	146	715	2026	4530

Source: Report on Currency and Finance, RBI, various issues, Economic Survey, Government of Orissa, Proceedings of SLBC, Orissa.

including cash crops and awareness of insurance products may be probable reason for higher premium paid by farmers of Andhra Pradesh. It is doubtful if the farmers in the State are aware of crop insurance and opt for insurance protection. As per NSS Report No. 496 on "Some Aspects of Farming, 2003", only 4 per cent of farmer households reported ever having insured their crops. Among those who have never insured their crops, it is seen that 57 per cent of farmers were totally unaware of practice of crop insurance. It was further found that 57 per cent of farmers were totally unaware of the practice of crop insurance, 16 per cent aware but not interested, 24 per cent said that the facility was not available to them and 3 per cent observed they could not afford to pay the crop insurance premium. Some land owners misuse this coverage treating insurance premium as the input and receipt of indemnity as the output.

Technology Transfer

The scope of technology transfer and extension activities related to agricultural engineering and technology is yet to be explored in a significant manner. Decline in public investment in agriculture in the post-reform period both at the national and the State level has adversely affected the prospects of capital formation and technology transfer to agriculture. This has been corroborated by the Bhattacharya Committee (2003) which points to the decline in Gross Fixed Capital Formation from 7.7 per cent of GDP in 1980-81 to 5.4 per cent in 1990-91 and to 3.9 per cent in 2001-02. It may also be mentioned that investment in agriculture as a per cent of GDP has also declined from 1.92 in 1990-91 to 1.27 in 2002-03 (Mishra, 2005). Such findings naturally call for a rethinking on State's role in agricultural development. The State, apart from investments in rural infrastructure may consider extending subsidized financial support to the enterprising farm households to undertake the production of fruits and floriculture. Along with the capital support to agriculture, regular training may be organized for the farmers on selection of seeds, implements and machinery for use in agricultural operations to raise the level of productivity for farms. The State has Agro Services Centres functioning in different regions. They serve as demonstration centres for the entrepreneurs in agriculture. Empowerment of these centres with technical, financial and marketing support will definitely add to the much needed crop diversification and create better income opportunities for the farming community. We are yet to see potential outcome of synergy between agriculture and these centres.

Diversification

To strengthen the development of agriculture in the State, there is a need to shift from the cultivation of traditional crops, mainly rice and go for diversification with the cultivation of pulses, oil seeds, cash crops and vegetables which have an emerging market and export value so that farmers get better net returns. Cultivation of fruits and vegetables by the farmers may hold potential benefits provided proper marketing and storage facilities have been created. Then only there will be consistent scope for further development. Farmers in the State may also be motivated to undertake the cultivation of medicinal plants (i.e., Kalmegh, Chicory, Isabgol) with scientific management practices for increasing the prospects of higher earnings for the farm households.

Odisha today enjoys unprecedented opportunities for growth in agriculture. The farm sector, to remain competitive, has to improve its operational efficiencies. Its long term prospects of growth remain encouraging if we take care of the different infrastructure deficits in the farm sector and address the following problems of the farmers:

Irrigation Facility: Agriculture in Odisha is mainly rain fed. Around 31 per cent of our agricultural land is irrigated compared to 92 per cent in Punjab and 73 per cent in Haryana. Yet irrigation facilities in the State are not spread uniformly. Efforts need to be made to expand the existing irrigation facilities by renovating and deepening of waterbodies. The State may even try to facilitate the development of tube, drip and fountain irrigation projects in areas depending on the type of crops and vegetables cultivated there. This may help us with the round the year cultivation of crops by providing assured water supply for the growth of the plants.

Supply of Electricity: Agriculture being a time bound activity, uninterrupted supply of power to the farm sector to enable the farmers to complete their diverse farm operations in time assumes significance. Failure to operate lift irrigation points and to carry on with important post-harvest activities because of the non-availability of power will cause dislocations in farm operations and thereby cause loss to the farmers. It will serve the purpose better if there is a dedicated line for the supply of power to agriculture.

Provision of Seeds and Farm Inputs: Successful cultivation of crops and vegetables would depend on the availability of quality seeds along with timely supply of chemical inputs. Farm households should have easy access to the institutions and sources providing high quality seeds and chemical fertilizers and pesticides for growing healthy and disease free crops and vegetables. This also calls for training and awareness campaigns among the farmers to remove their ignorance and illiteracy relating to the use of improved seeds and application of chemical fertilizers and pesticides.

Marketing and Storage: Increased production of cereals and fruits and vegetables would call for expansion of marketing facilities and creation of storage facilities. Marketing facilities, in fact, have to be made available to the growers at their doorsteps by making necessary buy back arrangements with them by the marketing agencies or the wholesalers. Cold storage facilities may also be created at the Panchayat/the Block level. This would take care of a long felt deficiency in the storage and marketing of crops and vegetables. A common complaint that has been voiced by the farmers in Odisha relates to unremunerative procurement price for rice, which is the principal crop produced in the State. Since the procurement price is fixed by the Union Government on the basis of recommendation of the Agricultural Prices Commission, the State government, as an incentive can announce a bonus above the procurement price for the producers in the State. The bonus is borne by the State.

Credit Supply and Insurance Cover: An important facility that needs to be made available in the farm sector is the provision of credit and insurance cover for the farm households. In the context of modernization of agriculture in the post-reform period, credit for farm operations becomes a critical input. It is the institutional sources of credit that has to cater to the growing credit requirements of the farm sector. It is important to note that Commercial banks, Cooperatives and Regional Rural Banks (RRBs) have become important sources of credit for the farm households. It may also be noted that farm finance from Commercial Banks, *Gramya* and Cooperative Banks stood at Rs.1,15,242 crore in 2004-05 (Rout, 2005). As commercialization in agriculture is taking place at a rapid rate, the Regional Rural Banks and the Cooperatives, Commercial banks should come forward to play a bigger role in extending finance to the farm households to enable them to develop horticulture and undertake cultivation of new crops and vegetables in response to the emerging tastes and preferences of the neo-rich consumers.

Insurance cover for the farm households in the State is virtually absent. Though the farmers in the State get compensation for their loss of crops due to calamities, comprehensive crop insurance cover needs to be worked out to provide meaningful bail out to the farm households, particularly to those farmers who have been trying to modernize their agriculture and venturing into new production areas.

Shortage of Wage Labour: Agriculture in Odisha, which once was a labour surplus sector, of late has been facing the problem of labour shortage. A commonly raised allegation refers to two-rupee rice scheme launched by the State to provide food security to the poor households which has adversely affected the supply of labour willing to work for a wage. It is also true that young labour from rural areas migrate to towns and cities to earn more. It is difficult to stop such migration. Yet a strategy may be worked out to retain rural labour by raising the wage rate and, if possible, by increasing the man days of employment under MGNREGA.

Crop Damage and Destruction: The threat which the farmers face in protecting their crops and vegetables from the monkeys, elephant and cattle need to be taken care of by the Forest Department of the State government. Farmers should receive protection from these animals, particularly at a time when the crops are ripe for collection by the farmers failing which adequate compensation is paid to them for the loss and damage to their fields caused by the wild animals.

Development of agriculture in the State offers a great challenge to all of us. The farmers in the State work amidst all the challenges.

Conclusion

Development of agriculture in Odisha needs to be tackled in an integrated manner by looking into all the aspects. Right from the ploughing of the fields till the crops are harvested and stored/sold, each stage forms an integrated part of the whole process. Not only that, Government agencies including the NGOs working in the field have to take the lead in bringing the scientific and technological breakthroughs in agriculture to the knowledge of the farmers and facilitate their implementation by creating enough confidence in the minds of the farmers who are generally shy of undertaking risks and uncertainties. It is also equally important that at each stage of agricultural operations the field staff of the government should be there to guide and assist the farmers. Given the need for agriculture development, visible presence of the State officials with a set of right policies, an improved infrastructure and entrepreneurial involvement of the farmers will make process of agriculture development more inclusive. Faith in a liberal market economy has to be blended with State policies.

Development in Odisha's agriculture will not come the easy way. All the stakeholders including the State have to be involved. Hard work in all fronts has to be put in. The result will be evident in the greater prosperity of agriculture and in the well-being of the farmers.

REFERENCES

Akthar, S.M. Jawed and Abdul Azeez (2012). "NREGA and Poverty Reduction in India" in Bhabesh Sen and Mitali Chinara (Eds.), *Economic Development and Poverty in India,* New Century Publications, New Delhi.

Behera, A.K (2005). "Challenges, Policies and Strategies for Agricultural Development in Orissa", *Journal of Indian Institute of Public Administration,* Bhubaneswar.

Chakrabarty, K.C (2011). "Prospects for Economic Growth and the Policy Imperatives for India", *RBI Monthly Bulletin,* January.

Government of Odisha, *Odisha Agriculture Statistics*, 2009-10, *Economic Survey*, 2010-11

Government of India, *Economic Survey* 2010-11

Mishra, Baidyanath (2005). "Post-Reform Trends in India's Agriculture Growth", *Journal of Indian Institute of Public Administration,* Bhubaneswar.

Naik, Dibakar (2005). "Agriculture Performance in Orissa: An Economic Analysis" *Journal of Indian Institute of Public Administration*, Bhubaneswar.

Panda, Prasanta Chandra & C. K. Parhee, (2005). Agricultural Insurance : An Inter-State Comparison, *Journal of Indian Institute of Public Administration*, Bhubaneswar.

Panda, Raj Kishore, (2005). "Agriculture Credit in Orissa : Developments and Constraints", *Journal of Indian Institute of Public Administration*, Bhubaneswar.

Panda, Sudhakar, (2012). "Growth and Poverty: An Assessment", in Bhabesh Sen and Mitali Chinara (Eds.). *op. cit.*

Rout, Jogesh (2005). "Agriculture in India And Orissa" *Journal of Indian Institute of Public Administration*, Bhubaneswar.

Section IV
Small Farmers and Food Policy

11

Nutritional Security through Diversified Food Production

Rita Singh Raghuvanshi

Introduction

India has emerged as a strong economy that was predicted last year to overtake China within 3 years by the Economist magazine. Significant progress has been made in food production in India over the years, with a record 241.56 million tonnes in 2010-11 (Ramadhar, 2011). Improvements have also been registered in other crop as well. As we know that Green revolution has transferred the country from food deficient State to food sufficient nation. During 2008, India ranked second in the world with regard to production of wheat (79 m. tonnes), paddy (148 m. tonnes), groundnut (in shell 7 m. tonnes), sugarcane (348 m. tonnes), vegetables and melons (79 m. tonnes), fruits excluding melons (6.3 m. tonnes) and first in total pulses (15 m. tonnes) (Boopathi, 2011). It is clear that India has attained self-sufficiency in foodgrain production.

Food Insecure Indian Population

Despite strong economic growth and achievement of self sufficiency in foodgrains at the national level, India is still home of 252 million undernourished people (Swaminathan, 2011). Hence, there has to be paradigm shift from emphasis on food security at macro level to nutrition security at the individual level. The issues and challenges for nutrition security are availability of food, i.e., production and access to food in right quantities or proportion. These challenges are to reinforce healthy eating practices, support nutritionally vulnerable population, and continue to enhance the availability of food and diversify food production.

Food security in India has recently been more alarming than ever before. Increasing pressure on land, environmental degradation and more frequent occurrence of natural calamities like flood, drought, etc., has worsened the scenario. Settled agriculture does not provide them sufficient food and health security.

Food security and nutrition security are not synonymous terms. However, nutrition security encompasses with food security. As defined by FAO "Food security is physical availability and economic access to sufficient, safe and nutritious food by all people at all times in quantities sufficient to meet their needs for an active and healthy life."

According to M.S. Swaminathan '"Nutrition security is defined as" physical; economic, environmental and social access to balanced diet and clean drinking water for every child, woman and man". In nutshell, food security is measured by agricultural produce and nutrition security as how the agriculture produce has been utilized by the biological system and given right parameters/indicators for human health. India has achieved food security by agriculture produce. But nutrition security is farfetched goal because a large number of people are malnourished.

Women Farmer and Nutrition Security

Though India has attained self-sufficiency in foodgrain production, about 48 per cent of the farm households are indebted. There is considerable variation in productivity levels of various districts across and within the States. Low levels of productivity in 51 per cent of districts will greatly affect the standard of living of farm households (both men and women), especially that of small farmers (Boopathi, 2011). India is a land of small farmers, with 650 million of her 1 billion people living on the land and 80 per cent farmers owing less than 2 ha of land. In other words, the land provides livelihood security for 65 per cent of the people, and the small farmers provide food security for 1 billion (zcommunication.org cited on 29-3-2012). Interestingly a large share of marginal farmers is women, particularly in hills or where men have migrated for greener pasture. Women's role particularly farm women's role in the economy has often been underestimated and their work in agriculture and providing food and nutrition security in the households has long been invisible. Women produce between 60-80 per cent of the food in most developing countries and are responsible for half of the world's food production, yet their key role as food producers and providers and their critical contribution to household food security is hardly recognized.

Women play diverse roles in attainment of food and nutrition secure family such as: major role in crop production, household animal production enterprises such as predominant role in herding, providing water and food, cleaning stalls and milking, processing of animal products particularly milk products and responsible for marketing. They are responsible for fishing in shallow waters and in coastal regions, producing secondary crops, gathering food and fuel wood, processing, storing and preparing family food and fetching water for the family. Women play important roles in food security as food producers, keepers of traditional knowledge and preservers of biodiversity, food processors and food providers for their families. Because of their multiple roles women are key players in overcoming food insecurity. Women tend to preserve traditional knowledge of indigenous plants and seeds. They have special knowledge of the value and diverse uses of plant parts for nutrition, health and income. They are involved in the exchange and saving of the seeds, growing of traditional varieties of vegetable and herbs. This has important implication for the conservation of plant genetic materials. Women are globally responsible for food preparation for their families and engaged in various stages and steps of processing the food. In many cultures and countries, women have the main responsibility for the provision of food if not by producing it, then by earning income to processing it. Development planners often ignore the gender division of social responsibility; false assumptions may not be fruitful to check food insecurity.

Role of Farm Women in achieving Nutrition Security

Women play a crucial role in curbing the crisis of malnutrition, food and nutrition insecurity. As guardians of food security for the family and principal care givers of young children, women are key partners in efforts to improve child nutrition security. Growing evidence shows that income in the hands of women contributes more to household food security, sustainable nutrition security for child and each member of the family by acting as food producers, than the income controlled by men. From the income available in the family, women in the households tend to allocate more of their budget to basic goods including nutritious food for themselves, children and all members of the family as compared to men who spend a substantial portion on entertainment for themselves only. Despite lower incomes and additional demands on their time as housewives and mothers, female headed households have the largest per cent of food security and a sustainable nutrition security. By acknowledging the role of women in nutrition; a vital link can be established between food securities and strengthening of sustainable family nutrition, health and care.

Major Nutritional Problem in India

In spite of large food production, malnutrition is rampant in India. It is in fact a public health problem. There is an absolute increase in undernourished persons from 210 million to 252 million (FAO, 2009). The National Family Health Survey (NFHS), 2005-06, highlights some very disturbing truths about the prevailing situation in India. Approximate 56 per cent of women are anaemic, 22 per cent of new born babies are of low birth weight (LBW); and 43 per cent children are underweight. Although it is widely reported that 20 per cent Indians are undernourished, 43 per cent children below five years are underweight, 70 per cent children and 55 per cent women are anaemic and 7.6 per cent is child mortality (www.nfhsindia.org cited on 2012). So, the country presents a contrasting picture of one-fifth population chronically hungry at lower strata even after being a food sufficient nation. This imbalance can be corrected. The Food Security Bill will be meaningful only when corrective measures are implemented through legislation on food wastage and safe storage, besides measures on enhanced food production.

There are serious deficiencies in the diets of our population particularly among the poor. As a consequence dietary deficiencies, several nutritional deficiencies with clinical manifestations and disabilities are encountered in our country, namely, (i) protein energy malnutrition (PEM), (ii) Vitamin A deficiency, (iii) iron deficiency anaemia, and (iv) iodine deficiency. PEM and vitamin A deficiency occur mostly among preschool children. Anaemia is prevalent in all groups, highest prevalence being among pregnant women, preschool children and women during reproductive age. In addition, India is heading towards overweight, obesity and non-communicable degenerative diseases.

(i) Protein Energy Malnutrition: PEM continues to be a major public health problem in many developing countries including India. PEM occurs mainly among young children of age due to improper introduction of weaning and supplementary foods. Stunting is present in 45 per cent children and wasting in 23 per cent children and underweight in 40 per cent children below age 3. About 33 per cent women and 28 per cent men are undernourished.

(ii) Vitamin A Deficiency: Vitamin A deficiency has been also recognised as a major public health problem among young children in India and also in many developing countries. Prevalence rates are higher in children below 3 years of age. A high proportion of children are going blind due to severe Vitamin A deficiency and die subsequently. Vitamin A supplementation can reduce child mortality. National averages show a declining variation exists between States and even within States. In backward areas Vitamin A deficiency is still a significant problem. India still has largest number of Vitamin A deficient children in the world. It precipitates more than 3 lack child death every year.

(iii) Nutritional Anaemia: Iron deficiency anaemia is the most widespread nutritional problem in the world. Although iron deficiency affects both genders, it is more common in women of reproductive age and young children. In India, iron deficiency anaemia affects more than half of the total population. Pregnant women and young children are affected the most. Children below the age of three years had a higher prevalence as compared to peers between the ages of 3-5 years. Estimates support that about 25 to 50 per cent girls become anaemic by the time they reach menarche and both rural and urban girls are similarly affected. Anaemia is perhaps the most common complication during pregnancy in developing countries and in India.

(iv) Iodine Deficiency Disorders (IDD): Iodine deficiency is the most common preventable cause of mental deficiency in the world today. The most familiar effect of iodine deficiency is goitre and cretinism. Goitre and cretinism is only the tip of the IDD iceberg. IDD refers to broad spectrum of conditions that affect the foetus, neonate, child, adolescent and adults. Iodine deficiency disorders may be reversible in early signs of goitre and hypo-thyroidism; and its irreversible manifestations are still births, abortion and endemic cretinism (Sehgal and Raghuvanshi, 2007)

(v) Overweight and Obesity: Obesity is defined as an abnormal growth of adipose tissue due to an enlargement of fat cell size or an increase in fat cell number or a combination of both. More technically body weight of more than 20 per cent of ideal body weight is considered obesity. Overweight is condition, where the body weight becomes 10-20 per cent more than ideal body weight.. Evidences from sixth century B.C.

show that an Indian surgeon Sushruta, related obesity to diabetes and heart disorder. He also recommended physical work to help cure it and its side effects.

Obesity is a chronic disease that is increasing in prevalence and contributes to chronic diseases such as cardiovascular diseases, diabetes, gall stones and arthritis. Obesity may lead to lowered fertility and at a lower side it may cause flat feet. Finally it leads to premature death. World Health Organization (WHO) in 1998, designated obesity as a global epidemic rate. Globally earlier obesity was expected only in developed countries like USA, UK, etc. But now obesity is escalating wildly along with malnutrition in developing countries as well. Developing world now faces the double burden of undernutrition as well as obesity. Prevalence of obesity in India is staggering and alarming. It is growing both in scope and pace. Various reports claim that at least 70 million people in India are overweight and more than 30 million are obese. In India obesity has reached an epidemic proportion affecting 5-7 per cent of the country population. In a study conducted in Uttarakhand, University girl students revealed that 29.73 per cent girls are undernourished and 16.06 per cent girls are obese and overweight. Various studies also show that childhood and adolescent obesity is also on increase now. The high incidence of obesity among adults in India is a major concern because people in age group of 35-55 years constitute major bulk of the countries productive group. Increase in incidence of obesity and other diet related degenerative diseases might cause great hazard on their health status and hence on their working efficiency.

Dietary Diversity: Need of the Hour: There are six nutrients, viz., water, protein, carbohydrates, fats, minerals and vitamins. Then there are various types of proteins, carbohydrates, fats minerals and vitamins. No single food crop can provide all the required nutrients in desired quantities

Table 11.1: Nutritional Status of Young Adult Females of Uttarakhand

BMI	*Age Range in years*					
	18-23 n=329		*23-28 n=300*		*18-28 N=629 Total*	
	n	*(%)*	*n*	*(%)*	*n*	*(%)*
<18.5	117	35.56	70	23.33	187	29.73
18.5-23	169	51.37	172	57.33	341	54.21
23-25	22	6.69	33	11.00	55	8.74
25-30	21	6.38	21	7.00	42	6.68
30-35	0	0.00	4	1.33	4	0.64

at a time. Therefore, it is important and essential that we consume diversified food items and ingredients in various recipes. It is essential to emphasize that only consumption of energy may take care of hunger but it shall not bring about nutrition security. For household nutrition security the challenges are:

- To reinforce healthy eating practices, support nutritionally vulnerable population, and continue to enhance the availability of food that support healthy eating. (Nutrition education in schools, nutritional awareness programme through multimedia).
- Diversify food production and formulation in adequate quantities for different season (Production of minor crops, fruits and vegetables, better storage, value addition and post-harvest processing are the issues).
- Implement policies and incentives in publicly funded organizations to promote increased availability of foods that support healthy eating. (ICDS, MDM, PDS, MNERGA)
- Maintain high standard of food safety through food regulation, inspection, quality control and consumer safe food handling practices and support nutrition research. (Quality of raw processed and cooked food through vendors, food canteens, fast food joints, restaurants).

The concerns for nutrition security are dovetailed with renewed emphasis on pro-poor agricultural policies in India. Focus on agriculture is important for nutrition security, because agriculture is a source of food, source of income, agriculture policies and food prices affect purchasing power and consumption pattern. In the agriculture and nutrition pathway gender dimensions to say women play an important role. Women's nutritional status, her ability to take care of young child and family members and her status in family, intra-household discussion and resource allocation influence nutrition security to a great extent.

Dietary diversity is essential for health and biodiversity is essential for dietary diversity. A diverse diet includes fruits, vegetables, milk, grains, coarse cereals, pulses and oilseeds which contributes to good health. Most essential nutrient deficiencies can be eliminated by small increases in the variety of food consumed. Many underutilized species, like beta carotene rich palm fruits are rich in nutrients. Millet is a good source of calcium and iron, the recent decline in the consumption of millet can be associated with an increase in iron-deficiency anaemia. The diverse nutrition and health functions that plants serve in traditional culture, and indigenous knowledge of plant diversity, offer potentially valuable solutions that enable biodiversity to address the problems facing contemporary society (www.ajol.info cited on 3-10-2012).

Nutrition Security and Biodiversity

Biodiversity often has a low priority on the development agenda. Nutritionists emphasize the importance of deficiencies in micronutrients such as iron, Vitamin A, iodine and zinc—so-called hidden hunger—to diet quality and disease resistance. In developing countries supplements and food fortification dominate nutrition programmes. Common treatments include extra doses of Vitamin A and fortifying staple foods such as wheat, maize flour or sugar with iron to prevent anaemia. While very effective in many cases, these measures are difficult to sustain. As the growing dependence of population worldwide on a few staple crops leads to increasing health problems, more attention is being given to biodiversity. The overall aim is to promote the sustainable use of biodiversity in programme contributing to food security and human nutrition, and thereby raise awareness of the importance of this link for sustainable development (www.who.in cited on 3-10-2011).

Underutilized Crops: Basket of Balance Diet

A balanced diet is one which provides all the nutrients in required amounts and proper proportions. It can easily be achieved through a blend of all basic food groups. The quantities of foods needed to meet the nutrient requirements vary with age, gender, physical activity and physiological status. A balanced diet should provide other non-nutrients such as dietary fibre, antioxidants and phytochemicals which bestow positive health benefits. Antioxidants such as Vitamin C and E, beta-carotene, riboflavin and selenium protect the human body from free radical damage. Other phytochemicals such as polyphenlos, flavones, etc., also afford protection against oxidative damage. Spices like turmeric, ginger, garlic, cumin and cloves are rich in antioxidants (NIN, 1998).

Underutilized crop species are often viewed as irrelevant and uneconomic solutions to global nutrition problems. However, underutilized crops do make essential contributions to adequate diets. Millets and pseudo cereals such as finger millet, buckwheat, *amaranth* are source of good quality protein, calcium, iron and fibre. Horsegram, cowpea, rice bean and *rajma* are grown locally and have good protein along with therapeutic uses. Fruits, minor vegetables, sauce ingredients, condiments, spices and medicines grown in small quantities are inexpensive and healthy complements to diets. Underutilized crops make important contributions to the nutrition and health of people in developing countries. Whereas these potential benefits provide a good reason for conserving agro biodiversity, this link receives little attention nationally and internationally.

A growing number of medical studies demonstrate that optimal health requires more than just essential nutrients. Such findings emphasize the potential value of underutilized species. Foods derived from buckwheat and finger millet, for example, reduce the risk of heart disease while bitter gourd and fenugreek contain compounds that directly improve the body's ability to respond to insulin. The benefits of leafy vegetables and other plants containing carotenoids such as lycopene and lutein are also well recognized. These carotenoids have no nutrient value but they do act as antioxidants and help prevent damage to cells and tissues. So, there is need to adopt a holistic approach to securing the sustainability of biodiversity, health and agriculture.

Wild Foods: Food Security Source in All Seasons

Wild and uncultivated biodiversity is ignored in dietary surveys, laboratory analyses of food composition in policy and decision-making. The widespread availability of high-energy stable foods such as rice, wheat, edible oil and sugar at relatively low cost has contributed to a double health burden in developing countries. Nutrient deficiencies increasingly co-exist with obesity and diseases such as diabetes and heart disease in many parts of the world. That's why attention is increased towards the importance of agro biodiversity.

As the poverty is prevalent in rural areas, these wild foods supplement their diet with almost all kinds of foods round the year. In this way, wild foods lead to best utilization of the natural resources. It helps in maintenance of biological diversity, biotic integrity and ecological processes forever.

Urgent Need of increasing awareness regarding underutilized and wild foods utilization:

(i) Intimate association exist between local underutilized crops, wild foods and various religious/cultural ceremonies. But, due to ignorance or lack of awareness regarding their nutrition and health benefits, these are decreasing in number. Now, it is right time to disseminate the information regarding their potential nutrition value and conserve them for future generation's sound nutrition and health.

(ii) Many of the uncommon crops as well as wild foods existence are on the verge of end due to the very fast adoption of western style of dietary habits. So, there is great need to save these endangered plants.

(iii) There is urgent need of extensive research work on these diversified local foods for production, consumption and utilization to make community nutrition secure in future.

(iv) Promote understanding among rural and local people regarding the linkage between biodiversity and sustainable nutrition security.

Ways for Improvement in the Present Situation

(i) **Fine-Tuning of Food with Nutrition Security:** Nutrition is the fundamental basic requirement for good health, functional efficiency and productivity. For a nation to be healthy, strong and productive, the nutritional status of its people must be good. Balanced farming means a nation cultivate not only abundant cereals but also pulses, oilseeds, more so fruits and vegetables. Market forces distort the balance in farming and require regulatory support from State. Pulses and oilseeds production have priority over cereal production in irrigated areas and continued mission mode approach would yield a desirable future (Vasimalai, 2011).

(ii) **Increase Diversified Food Production:** It is a cause of concern, sufficient food production and availability has not percolated to individual households, therefore, it has resulted in food insecurity and poor nutritional states. Challenges are to increase diversified food production for nutrition security. This can be done by giving due to the production of coarse grains, i.e., amaranth, buckwheat, millets, hoarse gram, rice bean and maize, etc. In addition to local fruits and vegetables; roots and tubers; local herbs and spices, they are there since centuries as they have varied nutrients and photochemical in different amounts available in different seasons thus supported health and nutrition of the population.

(iii) **Small millets production and consumption**: Small millets production and consumption has drastically come down because of cereal consumption through public distribution system. Small millets otherwise called nutri-cereals are rich in minerals with slow release of carbohydrates. Expanding small millets production and consumption would ensure nutrition security with sound health. Agriculture scientists should pay attention for large cultivation of these items to meet the population demand of food for nutrition security through women and small farmers (Vasimalai, 2011).

This shall improve access to nutrient rich foods. We should also work out possibilities of decentralized procurement of foods under PDS. Inclusion of finger millet, pearl millet, sorghum, maize, buckwheat, amaranth, horsegram, cowpea, eggs, soybean and so on in the decentralized procurement basket offer an excellent variety and opportunity to provide locally acceptable nutritious food to the people who have bean using it traditionally. Further it shall mitigate the problem of transport and storage. Mixing of grains offers opportunities of bio fortification. At research level bio-fortification of staple grains with inclusion of micronutrient (iron, zinc, Vitamin A) by selection, breeding or biotechnological tools is a good step for nutrition security.

(iv) **Post Harvest Management Practices:** Apart from increase in quantity demand there would also change in consumption pattern with greater demand for fruits and vegetables, processed food with greater emphasis on food safety and quality. Greater emphasis must be given to post-harvest management practices (Boopathi, 2011).

(v) **Farming as Enterprise:** The agrarian problems mainly concern the marginal and small farmers who experience untold miseries. For them, the profitability of farming is a major objective. Farmer's satisfaction and confidence in the mechanism is most crucial for sustained production. This would help to balance the food production with profitability/prosperity of farming community. Farming should be developed as enterprise.

(vi) **Farmer's Exchange System:** One of the main reasons for some farmers to be poor with nutritional security is that those who grow staple food crops like rice, wheat may not have additional resources to buy secondary food products like vegetables due to escalating prices.

Conversely, those who cultivate secondary crops may face the problem of poor accessibility to staple food items. This results in food and nutritional insecurity of many farm families. This imbalance can be corrected through an effective Farmer's Exchange System, in which one set of farmers (cereals) exchange at least 10 per cent of their produce with growers of other crops (vegetables) in kind but not in cash. This would ensure balanced food basket for the families with women and children. The farmer's associations can work out the details to ensure the normal health of farmers for the task of food production (Rajgopal, 2011).

(vii) **Paradigm Shift Towards Agriculture Development than Agriculture Growth:** Agriculture is undergoing a dramatic change. Growing price rise of inputs and seasonal swing of output price challenge viability of crop production. Now farmers have to integrate value addition activity with their crop production to improve viability of their farming. Range of value addition: simple grading and packing, semi-processing, standardised processed food are possible in a centralised manner. Structurally moving from farming to enterprise mode, both at household or farm and collective with different form including Producer Company with village based producer group, would be future way.

(viii) **Safe Storage at Village Level**: Safe storage at village is going to give an answer to price fall during harvest season.

(ix) **Cold Chain System:** Cold chain could also be created for fruits and vegetables to fetch a good price with quality.

(x) **Bankers Support:** Farmers collective with bankers support could organise safe storage at village and towns.

Women's groups, including SHGs under National rural livelihood mission, can become instrumental in meaningful convergence of health, nutrition, education, and other broad-based scheme addressing the deep rooted causes of under-nutrition. Examples of such group-centric pro-nutrition approaches include producing and consuming nutrient-rich foods through homestead horticulture and poultry interventions; establishing and maintaining micronutrient food fortification units; producing and marketing low cost, nutrient dense supplementary foods; developing primary food processing; nutrition services; and their children to access essential health and nutrition services; and catalyzing critical behaviour change for optimal health and nutrition outcomes in the long run through community mobilization including the involvement of *Panchayati Raj* Institutions around nutrition- specific issues and action.

Conclusion

It is well known fact that biodiversity and nutrition security are interconnected elements, therefore, diversified food production and consumption should be encouraged. For making our citizens of nation all round nutrition secure, it is our great responsibility to provide them diversified foods and help them to understand the concept of a biodiversity and sustainable food chain. Women, generally the small farmers play a major role in preserving seeds and growing local crops, therefore, they should be informed and educated about the importance of work they are performing. Increasing prevalence of micro-nutrient deficiencies has forced to modulate our plans, actions and evaluation processes designed to combat such deficiencies. Dietary modification with the help of underutilized and wild foods is considered the safest and most sustainable long term measure to control deficiencies. As threats to biological diversity accelerate, the plant genetic resources that guarantee the current and future production of healthy foods, beverages and medicines must be better used to improve the well-being of those whose food security and health is at risk.

REFERENCES

Boopathi, M. (2011). New Approaches Are Imperative. *The Hindu Survey of Indian Agriculture*. p. 64.

Gopalan, C., B. V. Ramasastri and S. C. Balasubramanium (2004). *Nutritive Value of Indian Foods*, Hyderabad, National Institute of Nutrition. ICMR. pp. 29-30.

Khadiyal, S.; P. K. Joshi and S. Mahendradev (2011). Strengthening the role of agriculture for a nutrition secure India IFPRI, Policy Note December 2011.

NIN. (1998). *Dietary Guidelines for Indians: A Manual*, National Institute of Nutrition, ICMR, Hyderabad-500007.

Rajgopal, V. (2011). Impact analysis needed. *The Hindu Survey of Indian Agriculture,* p. 24.

Ramadhar, (2011). What Ails Indian Agriculture. *The Hindu Survey of Indian Agriculture*, p. 64.

Sehgal, S. and R. S. Raghuvanshi (2011). *Major Nutritional Problems*. Textbook of Community Nutrition. ICAR. New Delhi. pp. 161-208.

Swaminathan, M. S. (2011). Grains mountains and hungry millions: The Indian Enigma. *The Hindu Survey of Indian Agriculture*. p. 7.

Vasimalai, M. P., (2011). Paradigm Shift Towards Agriculture Development, *The Hindu Survey of Indian Agriculture*, p. 52.

Verma, T. and R. S. Raghuvanshi (2001). *Combating Micronutrient Deficiencies: Anaemia and Vitamin A Deficiency*. AICRP, ICAR. Krishi Anusandhan Bhawan, Pusa, New Delhi.

Verma, T. and R. S. Raghuvanshi (2001). *Uncommon Plants Foods of India*. AICRP, ICAR. Krishi Anusandhan Bhawan, Pusa, New Delhi.

www.ajol.info cited on 3-10-2011.

www.agricultures network.org cited on 3-10-2011.

www.fao.org.cited on 3-10-2011.

www.nfhsindia.org cited on 26-7-2010.

www.zcommunication.org cited on 29-3-2012.

www.who.in cited on 3-10-2011.

12

Climate Change and Sustainable Agriculture: Finger Millet for Food Security and Nutrition

K.N. Bhatt

Introduction

Sustainable agriculture and reduction in poverty has been one of the most challenging issues for inclusive development in all developing regions. Degradation of agro-ecosystems deprives the poor of key productive resources and deprives the people whose livelihoods rely on utilization of agricultural resources. These potential linkages between widespread poverty and fragile ecosystems raise growing concerns on strategies for poverty reduction, equitable distribution of income and intergenerational equity (Shiferaw, *et al.,* 2009:258). Indian agriculture has experienced sharp decline over the years, particularly over the past two decades. While the contribution of agriculture including allied activities accounted for 13.9 per cent of Gross Domestic Products at 2004-5 prices in 2011-12 (agriculture alone only 12.3 per cent, forestry and logging 1.4 per cent and fishing 0.7 per cent) (*Economic Survey* 2011:9), about half of our population is either wholly or significantly dependant for their livelihoods on some form of farm activity, it may be crop agriculture, horticulture, animal husbandry or fisheries (Planning Commission, 2011: 88). Expansion of green revolution crops, mainly wheat and rice, has reached its peak with their dependence on heavy water doses, chemical fertilizers and pesticides leading to health hazards and contributing to climate change. We examine here an alternative solution to overcome warming stress, water stress and nutrition stress under the vastly changing climatic condition today for a sustainable agriculture in future.

Climate change is a long-term summary of average and the variable weather conditions. Human activities are mainly responsible for most of the warming and emission of Green House Gases today. As per Worldwatch Institute Report (WIR) (2009: xvi) 11 out of the last 12 years were noted among the 12 warmest years ever recorded in terms of global surface temperature. Melting glaciers, acidifying oceans and migrating species represent climate change clearly. IPCC Report 2007, clearly pointed out that the climate system was warmed by over 0.7 degree Celsius in the last 100 years and it is projected to increase by 1.6 to 6.9 degrees Celsius above average above 1750 preindustrial level resulting in sea level rise of 0.18–0.59 metres above by 2100. CO_2 concentration level of 278 ppm in 1750 pre-industrial state jumped over to 379 ppm in 2005 which further rose over to 382 ppm in 2008. At CO_2 concentration of 550 ppm 3 degree Celsius increase in warming is estimated. Consequently, foodgrain output is threatened by climate change if average temperature were to reach even 2.0 degrees Celsius above pre-industrial level (WIR 2009:13-15, 23). Climate change will be resulting in entirely new weather patterns which will lead to profound influence on agriculture at all scales. There is evidence that adaptability and resilience in production systems are becoming increasingly important to enable farmers to cope with climate change, increased climate variability and stressed environments amidst climatic extremes (Frison, *et al.,* 2011: 245-46).

The main focus of this chapter is to examine inter-linkages between incoming climate change and sustainable agriculture. Analyze Finger Millet (*Ragi*) as climate change compliant crop (CCCC) with ecological, food, biodiversity, nutrition and health securities. Explore feasibility of finger millet production, product diversification, markets, institutional alliances and its impact on the aspects of social and distributive justice. And, finally, search for how finger millet production can be initiated in rural India in association with various stakeholders.

Climate Friendly Food System and Sustainable Agriculture

Land-uses and food production systems are reported as the highest greenhouse gas emitters. In a climate-friendly production system, use of many available technologies and management practices could reduce the climate footprint of agriculture and land-uses and protect existing carbon sink in natural vegetation and soils. Initiatives in food and land-use sectors, saving standing forests and evolving sustainable food and forest systems will cool the planet (WIR, 2009, 48-49). India's National Action Plan on Climate Change, 2008 for mitigation and adaptation focuses on: Solar Energy, Energy Efficiency, Sustainable Habitat, Water, Sustaining Himalayan Ecosystem, Green India, **Sustainable Agriculture** and Sustainable Knowledge.

Out of 3,000 identified edible plant species only 10 annual cereal grains, legumes and oilseeds are grown in 80 per cent world cropland; half of it is covered by wheat, rice and maize. The farmers and communities are central actors who manage land, food and fiber while industry shapes incentives for choice of crops, quality standards and profitability (WIR, 2009: 37). Therefore, capacity building at local level is required to manage eco-agricultural landscapes and climate friendly land-use system. All agricultural intensification strategies today emphasize on the substitution and supplementation of ecosystem function by human labour and petrochemical products which contain the seed of destruction in terms of increased greenhouse gases, depleted water supplies and degraded soils. A production system to deliver intensification without simplification is required in order to bring in varietal mixtures, species intercropping and broader diversification strategies which ultimately improve productivity, enhance ecosystem functions and provide adaptability (Frison, *et al.,* 2011: 247).

Increased productivity, increased soil fertility, sustainable and secure food supply, resilient farm ecosystem, pests and diseases resistance and better nutrition and greater health are the results of agricultural biodiversity. It needs to be a central element of sustainable agricultural development. Unfortunately, modern agriculture reduces agricultural biodiversity. In fact agricultural production depends on a range of regulating and supporting ecosystem services which to a large extent are replaced in simplified agricultural systems by human supplied inputs. Landscape heterogeneity involving diversity of crop, livestock and agro-forestry elements at different scales is an important feature of many production systems. A UN report 2010 says that 925 million people suffer from chronic hunger in the world in addition to currently more overweight and obese than chronically hungry people. The most challenging task to food security today is improving yields of the major nutrients as around one billion people in the world still face starvation and one third global population suffers from micronutrient deficiencies often known as hidden hunger. In India improved use of minor millets among very poor farmers in some areas has shown multiple beneficial impacts on yields, incomes, profits, nutritional value, female empowerment, all leading to likely conservation of these crops and their biological diversity in agricultural fields (Frison, *et al.*, 2011: 238-245).

India's agricultural development has been generally considered to be uneven both in terms of regions and crops. Coarse cereals basic in satisfying hunger have been left behind by the green revolution. These cereals have long been the staple diet of peasants and labourers with far better nutritional value and playing important role in the livestock economy in drought-prone areas. Although the area under coarse crops got diverted to other crops with the introduction of irrigation, their yields at least in areas which cannot be shifted to superior cereals continuously increased almost close to that of foodgrains as a whole. During 1950-51 to 1984-85 the proportion of small millets declined sharply from 27 per cent to a mere 9 per cent of the total coarse cereals production, facing the risk of extinction and threatening the crop diversity in semi-arid areas. Any growth strategy with increasing reliance on the production of rice and wheat alone could be disastrous in several ways. It could put too much of a burden only on two crops making food production more vulnerable and unstable, put further concentration of irrigation and agricultural prosperity only to a few areas to the detriment of others and also lead to the extinction of our rich crop diversity (Nadkarni, 1986: A-113-118).

Rainfed agriculture primarily provides grain food in most countries of the world. These semi-arid tropics (SAT) are the home of 38 per cent of the developing countries poor, 75 per cent of whom live in rural areas; over 45 per cent of the world's hungry and more than 70 per cent of its malnourished children live in the SAT. However, a grim picture of natural resource scarcity, low rainwater use efficiency (35-45%), adverse socioeconomic factors, poor infrastructure and inappropriate policies have become the basic characteristics of these areas which cause extreme poverty among farming households today. A global assessment report shows that 57 per cent of the total area of dry lands in two major Asian countries China (178.9 million ha) and India (108.6 million ha) are degraded (Wani, *et al.*, 2009:1-5). Available evidence from a long term experiment within India demonstrated virtuous cycle of persistent yield increase through improved land, water and nutrient management in rainfed agriculture. A large potential of rainfed agriculture remains to be tapped. An Integrated genetic and natural resource management approach cares for various components of the rural economy: traditional foodgrains, new potential cash crops, livestock and fodder production, socio-economic factors. Rainfed agriculture among the resource-poor farmers of the world may be upgraded through another important system of conservation agriculture. A new beginning has been recently made in India with the establishment of National Rain-fed Areas Authority (NRAA) with the mandate to converge various programmes for integrated development of rainfed agriculture (Wani, *et al.*, 2009: 10-23).

Minor Millets the Crop of the Future

There are two types of millets, i.e., major millets and minor millets. Major millets including great millet/sorghum (*Jowar*) and pearl/Spiked millet (*Bajra*) are tall growing and fairly drought tolerant while minor millets are known as a group of grassy plants with short slender Culm and small grains possessing remarkable ability to survive under severe drought. The deep root system of minor millets with efficient soil moisture extraction ability makes them capable to grow under very low to severe moisture stress, where most other cereal crops may not survive to produce grain. These crops are known for their extreme hardiness, resilience, survive even in lower than 300 mm rainfall and can sustain in shallow, less fertile soils, steep slopes of hillocks and mountainous terrains (Bala Ravi:1-2). India is largest producer of small millets, often referred to as coarse cereals. Small millets grown in the country are finger millet (*Eleusine coracana*), common or proso millet (*Panicum miliaceum*), little millet (*Panicum sumatrense*), foxtail or Italian millet (*Setaria italica*), barnyard millet (*Echinochloa crusgalli and E.colona*), kodo millet (*Paspalum scrobiculatum*), which are cultivated in an area of about 2.7 million ha. These crops are grown right from sea level in coastal Andhra Pradesh to 8,000 feet above sea level in hills of Uttarakhand and North-Eastern States in different kinds of ecological niches where crop substitution is difficult. Therefore, it is important to enhance production and productivity of these crops to ensure food and nutritional security (Shanthakumar, *et al.*, 2010: 117).

The small seeded crops minor millets are grown for food and fodder around the world. All these species are characterized for resilience and ability to thrive in harsh environments with nutritious seed content. These grains are also described as nutritious millets and have received far less research and development attention than other crops with regard to crop improvement, cultivation practices and utilization. Minor millets account for less than one per cent of the foodgrains produced in the world today. They are considered less important in terms of the overall world food production. However, they are strategic in terms of their nutritional contribution and their crucial role in local agro-ecosystems. These crops are mostly grown in marginal areas or under agricultural conditions where major cereals would fail to give sustainable yields (GFU ND: 1). India produces numerous types of millets both major and minor millets. Among great millets Sorghum (*Jowar*) and Spiked Millet/Pearl Millet (*Bajra*) are mostly known to all. However, there exists real information asymmetry in case of miner millets: finger millet (*Ragi*), Foxtail Millet/Italian Millet (Kaun), Little Millet (*Sama, Sava*), Kodo Millet, Common Millet/Porso Millet (*Cheena, Cheeno*) and Barnyard Millet (*Shyama, Sanwa*). All millets as stated above were traditionally known for their high nutritive values, but finger millet (*Ragi/Madua*) may be considered as one of the best for its nutritional components, medicinal value and health benefits. We are presenting here the case of finger millet (*Eleusine coracana*) as crop of the future for sustainable agriculture in an era of severe climate changes.

Finger Millet as in case of other minor millets is also known by several names in different States. Its botanical name is *Eleusine Coracana.* In Hindi we call it *Ragi, Mandika, Marwa, Madwa.* In Bangali it is known as Marwa, in Gujrati as *Nagli/Bavto*, in Kannada as *Ragi*, Marathis call it *Nagli/Nachni*, in Oriya *Mandia,* in Punjabi *Mandhuka/Mandhal*, in Tamil *Keppal/Ragi, Kelvaragu*, and in Telugu it is termed as *Ragi Chodi.* Table 12.1 shows different names of minor millets in various States of India.

Finger Millet — A Miracle Grain

Finger millet is the sixth most important grain in the world. It sustains a third of the world's population. It is only surpassed by maize as a major cereal in the Eastern and Central Africa (ECA) region (Okwadi, 2007: 102). Finger millet accounts for 8 per cent of planted area and 11 per cent of production of all millets worldwide. Precise data on area under finger millet are not available, because it is frequently reported with other millets including pearl millet. However, some estimates suggest that it is grown on over 4 million ha globally (Oduori and Kanyenji, 2007:10) with an estimated production at 4.5 million metric tonnes per annum (Okwadi, 2007: 102). Its native home is presumed to be eastern/central Africa, specifically Uganda, from where it moved to the Indian subcontinent around 3000 BC. It is a hardy crop, able to thrive in poor soil and limited moisture conditions (Mgonja, *et al.*, 2007:

Table 12.1: Minor Millets: Different Names

Language	*Finger/ Birds-food/ Coracana/ African Millet*	*German/ Italian/ Foxtail Millet*	*Kodo/ Ditch Millet*	*Barnyard/ Japanese Millet*	*Little Millet*	*Common/ broomcorn/ Hog/ Hershey/ Porso millet*
Bengali	Marwa	Kaun	Kodo	Shama	Sama	Cheena
Gujarati	Nagi/ Bavto	Kang	Kodra	–	Gajro	Cheno
Hindi	Ragi/ Marwah	Kakun	Kodon	Sanwa	Kutki/ Shava	Cheena/ Barri
Kannada	Ragi	Navane	Harka	Oodalu	Sama/ Save	Baragu
Marathi	Nagli/ Nachni	Kang/ Rala	Kodra	–	Sava/ Halvi	Vari
Oriya	Mandika	Kanghu/ Kora	Kodua	Khira	Suan	China
Punjabi	Mandhuka	Kangni	Kodra	Swank	Swank	Cheena
Tamil	Keppal/ Ragi	Tenai	Varadu	Kathiraivolly	Samai	Pani Varagu
Telugu	Ragi Chodi	Korra	Arikelu/ Arika	Kodisama	Samalu	Variga

50). With its wide range of adaptability, the crop is cultivated from sea level in parts of Andhra Pradesh and Ṭamil Nadu in India to about 2,400 metres above sea level in hilly areas in northern India. Similarly, it grows at high altitudes in Nepal, Uganda, Kenya and Ethiopia. It can survive adverse soil and weather conditions better than most other foodgrain crops (Upadhyaya, *et al.*, 2007:90-91). It can grow with a rainfall varying from 350 to 800 mm under different temperature regimes. It is adapted to a range of soil types, but grows better on red, light black or grayish loamy and sandy soils. In areas where there is adequate rainfall during the crop growth phases, finger millet can grow better in poor soils and produce reasonable yields as compared to sorghum and pearl millet (Kisandu, *et al.*, 2007: 24)

The reported grain yield potential of finger millet ranges from 4,265 kg/ha in Uganda, 6,060 kg/ha in Zimbabwe, 3,700 kg/ha in Ethiopia and 4,789 kg/ha in India. In other countries including China, Democratic

Republic of Congo, Eritrea, Kenya, Myanmar, Nepal, Tanzania and Rwanda finger millet is grown in large areas. The ICRISAT gene bank at Patancheru, India, holds 5,949 finger millet accessions from 23 countries: Burundi, Pakistan, Cameroon, Senegal, Ethiopia, South Africa, Germany, Sri Lanka, India, Tanzania, Italy, Uganda, Kenya, UK, Malawi, USA, Maldives, Zaire, Mozambique, Zambia, Nepal, Zimbabwe, Nigeria. This collection was established to work out strategies to enhance its use in crop improvement. The accessions are classified into six botanical races: *Africana, Spontanea, Compacta, Elongata, Plana* and *Vulgaris* with a flowering range of 63.17 days (*Africana*) to 83.69 days (*Elongata*) (Upadhyaya, *et al.*, 2007: 90-95)

Finger millet is suitable as a subsistence food crop as it can be stored being small seeded safely for many years without insect damage. In drought prone areas where harvests frequently fail this is particularly valuable (Oduori and Kanyenji, 2007:11). The grain is highly nutritious. Protein content is about 7.4 per cent, which is comparable to rice. Calcium content in whole seed is 0.34 per çent compared to 0.01-0.06 per cent in most cereals. Iron content is also exceptionally high, 46 mg kg-1 much higher than wheat or rice (Upadhyaya, *et al.*, 2007:91). In addition to its importance as a staple food, finger millet contributes greatly to the incomes of many households. In Uganda it is brewed to give a local beer that is sold for money or used to pay for labour. All Ugandan urban and rural areas have a thriving traditional beer brewing industry. Finger millet has several other advantages. It is reputed to taste better than most cereals; has no major storage pest problems and can be stored cheaply for long periods, provided it is dried well to low moisture content. These attributes combine to make finger millet a suitable crop for ensuring food security in drought prone areas. (Wanyera, 2007: 3-4).

Major constraints in finger millet include agronomy, disease, low productivity, poor post-harvest systems and issues related to output markets (Silim, 2007: xii). The most serious biotic constraint to the crop is blast caused by the fungus *Magnaporthe grisea.* It can affect finger millet at all growth stages, causing large yield losses. Researchers from National Agricultural Research System (NARS) of Uganda and Kenya, the International Crops Research Institute for the Semi-Arid Tropics (ICRISAT) and the UK National Research Institute (NRI) collated available germplasm, collected finger millet blast isolates and studied the epidemiology of the blast fungus and identified blast resistant genotypes. On-farm trials were conducted in Kenya in 2005 to evaluate and promote the identified blast-resistant varieties. Seven improved varieties were tested. Farmers valued the improved varieties for their yield and blast resistance. The experiment emphasized on more investment in finger millet research, extension and market linkages for improved food and nutrition security and enhanced income. *Helminthosporium* is another disease causing leaf-spots, seedling blight, and head blight which can be controlled by uprooting and burning infected plants as well as use of resistant varieties. The pests attacking the crop include grasshoppers, armyworms, shoot fly and stem borers. A parasitic weed *Striga*, may severely affect the crop and birds can cause some low level damage at harvesting stage. Finger millet production is labour-intensive, especially for controlling weeds accounting approximately 25 per cent of the overall labour requirement (Mgonja, *et al.*, 2007: 49-52). The crop cannot tolerate flooding. It can tolerate cold but requires rainfall spread over the growing season. It appears, however, to be photoperiod sensitive, the optimum being 12 hours. Problems of photoperiod sensitivity were encountered in the crop which otherwise grew successfully in the USA as far north as Davis, California (Mackey, *et al.*, 2007: 86).

The neglect of finger millet has contributed to a lack of realization of the potential productivity of this crop. The constraints are in the whole food supply chain from production to utilization including productivity enhancement, especially varieties that are resistant to blast disease, commonly known as a rice disease, high labour demand for weeding, under-developed seed systems, poor post-harvest handling which reduces grain quality, product development and an inefficient production-supply chain (ICRISAT, 2007:158). The Eastern and Central Africa Regional Sorghum and Millet Network (ECARSAM) was initiated in 2003 with the basic aim of development of these crops (Mitaru, 2007: 153). Finger millet is reported to be roughly where wheat was in the 1890s in terms of plant breeding research. Its yields could rise similarly like wheat or much faster, because it is a C4 crop (wheat is a C3 photosynthesizer) and because it can benefit from research results and techniques developed on other crops. Finger millet has very wide diversity and variability that would facilitate breeding and selection. It is reported that over 2,500 accessions of finger millet have been collected in East and Southern Africa (Oduori and Kanyenji, 2007: 16).

Commenting on the qualities of finger millet (*Ragi*) as a food crop, India's Commission on Agricultural Costs and Prices (CACP) observed that: "*Ragi* is a very hardy crop as well as a grain of great nutritive value. *Ragi* is grown in most of the States under dry land conditions, mainly by small and marginal farmers. Once harvested, the seeds keep extremely well and are seldom attacked by insects/pests. The long storage capacity makes *Ragi* an important crop in risk avoidance strategies for poorer farming communities. This crop has the potential to improve nutrition, boost food security and foster rural development...The Commission recommends that in order to ensure food security, coarse cereals should be included as a component of ongoing National Food Security Mission" (CACP, 2010: 173-74). As noted above, *Ragi* can withstand three challenges from Climate change: (i) warming stress, (ii) water stress and (iii) nutrition stress. It may be considered as a future climate change compliant crop (CCCC).

Finger millet in fact as described above is a miracle Grain of rain-fed areas. Out of 141 Mha. net sown area in India, 85 Mha. (60%) is rain-fed in 177 districts. Rain-fed agriculture contributes about 44 per cent of total foodgrain production in the country. Rain-fed regions provide livelihood to nearly 50 per cent total rural workforce and sustain 60 per cent cattle population. Finger millet provides a holistic and bio-diverse farming system for rain-fed areas as several mixed crops could be grown with this miracle grain in the same farm including varieties of both pulses and oilseeds. In comparison to the rainfall levels varying between 2000 mm to 1300 mm for the production of sugarcane, banana and rice, only 350 to 400 mm rainfall (20%) is required for Finger millet production MINI, DDS, FIAN (ND: 4). Finger millet is adaptable to wide range of ecological conditions, as it is grown in all corners of India right from high Himalayas to coastal areas and it also grows even in saline soil. As per 2007 report of Vivekanand Hill Agriculture Research Institute (2007), productivity level of finger millet ranges between 10 to 20 Qtls./ha. while Hybrid varieties developed by the Institute may produce in the range of 20 to 35 Qtls/ha. It is a 90 – 100 days crop. The crop grows well only on farmyard manures and bio-fertilizers and requires no synthetic fertilizers. As the storage pests are virtually nonexistent in finger millet for over long periods, India's huge fertilizers and pesticides subsidies which benefit only about 20 per cent rich farmers may be drastically reduced with the expansion in the area under finger millet and its production.

Finger Millet Nutritive Value

Finger millet has high levels of calcium and iron and lysine, an amino acid limited in most cereals. In Tanzania it is known as 'Ulezi' which means 'bringing up or nurturing' thus emphasizing the role of finger millet as the food for infants, breastfeeding mothers and as food given to those who are recuperating from diabetics and other illness (Silim, 2007: xii). The main protein fraction *eleusinin* in finger millet has high biological value with high amounts of tryptophan, cystine, methionine and total aromatic amino acids phenylalanine and tyrosine. All these are crucial to human health and growth but deficient in most cereals. The two sulphur-containing amino acids, methionine and cystine, especially, are lacking in the diets of millions of the poor who live on starch foods. This makes finger millet an important preventive against malnutrition. Children from finger millet eating parts of Uganda suffer less from nutritional diseases compared to those from banana eating areas. Finger millet is also a rich source of calcium. Some samples contain 0.33 per cent calcium which is 5 to 30 times more than in most cereals. Similarly, the phosphorus and iron contents are also high (Wanyera, 2007: 3). It is quite superior to rice and wheat in minerals and vitamins containing 17.8 times more fibre than rice; 34.4 times more calcium than rice; and 20 times more iron than rice. Table 12.2 depicts comparative nutritive values of finger millet.

Table 12.2: Composition of Finger Millet, Rice and Wheat

Contents	*Ragi*	*Wheat*	*Rice*
Protein (%)	9.27	11.80	6.80
Fat (%)	1.35	1.50	0.50
Fibre (%)	**3.56**	**1.20**	**0.20**
Mineral Salt (%)	2.11	1.50	0.60
Carbohydrate (%)	74.73	71.20	78.20
Energy (Kcal/100 g.)	348.00	346.00	345.00
Thiamine (Mg. per 100 g.)	0.42	0.49	0.06
Riboflavin (Mg. per 100 g.)	0.19	0.17	0.06
Niacin (Mg. per 100 g.)	1.10	4.30	1.90
Zinc (Mg. per 100 g.)	3.30	2.20	1.30
Chromium (Mg. per 100 g.)	0.028	0.006	0.003
Manganese (Mg. per 100 g.)	5.49	2.29	0.51
Magnesium (Mg. per 100 g.)	137.00	132.00	64.00
Iron (Mg. per 100 g.)	**14.03**	**4.90**	**0.70**
Calcium (Mg. per 100 g.)	**344.00**	**48.00**	**10.00**
Potassium (Mg. per 100 g.)	408.00	315.00	–

Source: Goplan, G., B.B. Ramshastri and S.C. Balasubrahmanium (1999): Nutritive Value of Indian Foods, National Institute of Nutrition, Hyderabad, Cited in Srivastava, Sarita (2008): *Mandua Ke Paushtik Byanjan*, G.B. Pant University of Agriculture and Technology, Pantnagar, p. 44.

Finger millet produces excellent malt extract for brewing and other similar industries (Kisandu, *et al.*, 2007: 25) Though often known as a crop for the poor, it is fast becoming a popular food crop among people of all wealth categories. It is recommended as the ideal food for diabetics, elderly, the sick and especially those affected by AIDS. Finger millet's long term storage capacity makes it critical in ensuring food security for poor households in drought-prone areas (Mgonja, 2007: 51). Finger millet straw makes good animal fodder, containing up to 60 per cent total digestible nutrients. It grows faster than maize, producing a large tonnage per acre, and could be harvested much earlier than maize (Mackey, *et al.*, 2007: 86).

Finger Millet Health Solutions

Finger millet helps in treatment of measles, anemia and diabetes. It has industrial potential in the manufacture of baby foods, foods for sick persons, and breakfast cereals. It has good properties for Malting which is the process of germinating finger millet to activate enzymes that break down the complex starches into sugars and other simple carbohydrates easy to digest. The malting qualities of finger millet are considered second only to barley. While barley is a temperate crop, finger millet thrives in tropical areas where malnutrition is endemic. Finger millet could provide cheap and nutritious foods for preventing malnutrition in babies food with its malting qualities, which causes large number of deaths throughout the tropics. The germinated (malted) grain can be used to liquefy any starchy food to make cheap, easily digestible liquid foods for children (Oduori and Kanyenji, 2007: 18-19).

Finger millet is highly calcium rich for bone formation of children. Being very rich in fibres, it is known for its slimming solution qualities as it digests very slowly. This food contains Thiamine, Riboflavin, Chromium, Potassium, Zinc, Magnesium, Manganese which are required for treatment of diabetic patients are relatively high in finger millet in natural form. Finger millet soup is known for its qualities as pain killer. It maintains body temperature in winters. In Uganda finger millet is used in gift giving, as a symbol of fertility and manhood (Wanyera, 2007: 6-7)

Finger Millet Food Products

Finger millet is used to prepare various products in Kenya: fermented porridge, non-fermented porridge, known as *ugali* and seasoned thick porridge (made from a mixture of finger millet and legume flour). A significant amount of the grain is also consumed in the local brewing industry, making opaque beer. Flour can be used for baking bread and various other products with good flavour and aroma; several brands of finger millet flour produced by different companies are available in Kenyan supermarkets. Malted finger millet (sprouted seeds) is a nutritious food, easily digested, particularly recommended for infants and the elderly. The straw makes good fodder that contains up to 61 per cent total digestible nutrients, more nutritious than pearl millet, wheat or sorghum. The grain can also be used for poultry feed. The key issue is nutritional quality (Oduori and B. Kanyenji, 2007: 15-18). In India numerous finger millet food items are prepared like sprouted finger millet, sprouted *salad,* roasted seeds, *chapati*, baby food, *halua, barfi, laddoo, chikki*, cake, cake-rusk, biscuits, bread, *kachauri, cheela, bhelpuri, bati, poha, namkeens, sewai, dosa, idli.* These products are laboratory tested for their high nutritive quality by the Department of Foods and Nutrition, G.B. Pant University of Agriculture and Technology, Pantnagar over past several years (Srivastava, 2003, 2008).

Finger Millet Social and Distributive Justice

Finger millet in fact may be considered as the food and nutrition security of the poor. In India, majority of the farmers are marginal and small who mostly suffer from poverty and malnutrition. As per the report of a National Commission, constituting 84 per cent of all farmer households, marginal and small farmers are considered as the backbone of Indian Agriculture. These farmers own only 43 per cent land and operate about 46 per cent of area and produce half the output. They suffer from prevailing market failures in agriculture NCEUS (2008: 29). The traditional Indian *Jajmani* System gave birth to the system of *Inam* Lands which are mostly dry and least productive. These lands are suitable only for the production of dryland crops like millets. Earlier, Scheduled Castes and Scheduled Tribes and other backward communities mostly sustained on the production and consumption of millets. Gradually unfortunately, white and bright green revolution crops replaced the black. In order to provide direct benefits to the poorest with highly positive forward and backward linkages for development of the Indian economy, the need of the present is promotion of nutri and bio-diverse cereals millets' production. This is also essential for ecological restoration as finger millet may bring numerous environmental gains.

Depleting Finger Millet

Areas under finger millet cultivation declined by 2.64 per cent per annum during 1986-87 – 2007-08 as shown in Table 12.3. There has been almost no state support over several years for the promotion of the crop. It might disappear in next 50 years at this rate which would be a civilization and ecological disaster for the humanity. The solution comes from focus on finger millet farming systems, enacting suitable policies and creating enabling environment. India is still a largest producer and consumer of finger millet. Skills for its production are available in all over rural India.

Table12.3: All India Trends in Area, Production and Yield of Finger Millet and Rice

Area : '000 ha, Prod. : '000 tonnes, Yield : Kg./Ha

Particulars	*1986-87*		*2007-08*		*Compound Growth Rate (1986-87 – 2007-08)*	
	Ragi	*Rice*	*Ragi*	*Rice*	*Ragi*	*Rice*
Area	2397	41154	1366	43796	–2.64	0.30
Production	2586	60906	1983	93947	–1.26	2.09
Yield	1078	1480	1437	2145	1.38	1.78

Source: CACP, 2010, p. 457.

Growth rate of foodgrains production in India today has "implications for the macroeconomic stability given the rising demand of the 1.2 billion people for food. Achieving minimum agricultural growth is a prerequisite for inclusive growth, reduction of poverty, development of rural economy and enhancing of farm incomes." Unfortunately, growth rate index of total coarse cereals during 1980-81 to 2011-12 was negative demonstrating either shift to other crops or relatively dry area remaining fallow. However, the production and yield growth rates which were 0.40 per cent and 1.62 per cent in 1980s improved significantly to 3.01 per cent and 4.39 per cent respectively during the period 2000-01 to 2011-12. It shows growing popularity of coarse cereals as nutria-food (Ministry of Finance, 2012:179, 182).

Crafting Markets for Finger Millet

Crafting of markets for finger millet only will lead to distribution of benefits to the socially deprived. There could be two methods for the creation of markets for this wonder crop. One is policy changes at Central and State Governments' level. Finger millet foods mix may be introduced for Indian Defence Forces, Indian Railways, PSUs canteens, Mid-Day Meal, academic institutions, hostels, etc. Both Central and State governments may earmark substantial budgetary allocations for the development of all varieties of millets including finger millet. Secondly, awareness generation through multiple sources: media, spiritual leaders and reputed individuals may break information asymmetry about finger millet's excellent food, nutrition, medicinal, biodiversity and environment regeneration qualities. One example that may be cited here is the recent awareness campaign in India about the benefits of *Lauki and A*loevera leading to huge demand for both these products. Similar efforts may be made for popularizing millets. Several institutions of repute may collaborate in the development of finger millet including technology, genetics, high yielding and disease resistant varieties, market exploration, awareness generation, products development, etc. In fact, efforts may be made for creating India as a finger millet nation.

It is indeed a positive policy initiative to note that Finance Minister of India in the budget of 2011-12 made a special provision of Rs. 300 crore, among others, on finger millet as a Nutrient. India's 2011-12 budget clearly noted that: "While we ensure food for all, we must also promote balanced nutrition. *Bajra, jowar, ragi* and other millets are highly nutritious and are known to possess several medicinal properties. The availability and consumption of these Nutricereals is, however, low and has been steadily declining over recent years. A provision of Rs.300 crore is being made to promote higher production of these cereals, upgrade their processing technologies and create awareness regarding their health benefits. This initiative would provide market linked production support to ten lakh millet farmers in the arid and semi-arid regions of the country. The programme would be taken up in 1000 compact Blocks covering about 25,000 villages. This will help improve nutritional security and increase feed and fodder supply for livestock." First time after independence great nutritional value of millets got recognition and support for their development in the country.

This budgetary allocation was made under *Rashtriya Krishi Vikas Yojana* for promotion of millets as Nutri-cereals. A Scheme on Initiative for Nutrition Security through Intensive Millets Promotion has been subsequently formulated to operationalize the announcement. The scheme aims at improved production and post-harvest technologies of millets in an integrated manner within the country. The Scheme through processing and value addition techniques is expected to generate consumer demand for millet based food products besides increasing production of millets. The Scheme was implemented from *Kharif* 2011. It was proposed that the technology demonstrations in compact Blocks would be organized in selected districts for four categories of millets – sorghum, pearl millet, finger millet and small millets in the identified districts during 2011-12. *Panchayati Raj* Institutions are proposed to be involved in the selection of village units including the participating farmers.

Conclusion

Indian agriculture has experienced sharp decline, particularly over the past two decades. Expansion of green revolution crops, mainly wheat and rice, has reached its peak with their adverse environmental impacts. Therefore, an alternative solution to overcome today's problems is required under vastly changing climatic conditions for a sustainable and bio-diverse agriculture in future. Capacity building at local level is needed to manage eco-agricultural landscapes and climate friendly land-use system. A production system to deliver intensification without simplification may be put in place in order to bring in varietal mixtures, species intercropping and broader diversification strategies which ultimately improve productivity, providing better nutrition and greater health and enhance ecosystem functions and provide adaptability. In India improved use of minor millets among very poor farmers in some areas has shown multiple beneficial impacts on yields, incomes, profits, nutritional value, female empowerment, all leading to likely conservation of these crops and their biological diversity in agricultural fields. Coarse cereals which are more basic in satisfying hunger and meeting the needs of nutrition have been left behind by the green revolution. These cereals have long been the staple diet of peasants and labourers with far better nutritional value and have also been playing an important role in the livestock economy of drought-prone/rainfed areas.

Minor millets are characterized for resilience and ability to thrive in harsh environments with nutritious seed content. Unfortunately these nutritious food crops have received far less research and development attention than other crops with regard to crop improvement, cultivation practices and utilization. Finger millet is suitable as a subsistence food crop as it can be stored safely for many years without insect damage. In drought-prone areas where harvests frequently fail this is particularly valuable. With its wide range of adaptability, the crop is cultivated from sea level to about 8000 feet above sea level in hilly areas in India. It can grow well with a rainfall varying from 300 to 800 mm under different temperature regimes, adapted to a range of soil types, can tolerate alkalinity and acidity and produce reasonable yields. Finger millet has high levels of calcium, iron and lysine, an amino acid limited in most cereals. It is known as a food for 'bringing up or nurturing' of infants, breastfeeding mothers and as food given to those who are recuperating from illnesses like measles, anamia, diabetes and HYV AIDS. Finger millet is used to prepare various food products, making opaque beer and Malt for human consumption and its straw makes good fodder for animals, poultry feed. Promotion of finger millet production may bring in distributive and social justice to the poorest as most of the marginal and small formers in India possess dry lands where the crop may grow better given suitable policy incentives. It is also essential for ecological restoration as finger millet has excellent qualities as a climate change compliant crop.

All Millets including the wonder crop finger millet no doubt are known for their excellent qualities for providing food, nutrition, health and ecological securities. Finger millet may provide great socio-ecological bonus to farmers including climate change bonus, bio-diversity bonus, water conservation bonus, organic food bonus. As such higher production, upgraded processing technology and improved disease resistant seed varieties of finger millet will go a long way in bringing multiple benefits for the country. Finger millet food items may be introduced in public distribution system throughout India. Similarly finger millet foods may be introduced in Indian Defence forces, Railways, ICDS, Schools, Colleges and Hostels, baby foods, PSUs. The crop is grown almost in all parts of India and people have skills for its production. It is indeed an equity ensuring crop system. People centred and people directed R&D activities for development of this highly nutritive food system may bring in numerous environmental benefits for the country. Both public and private sector enterprises may initiate their corporate social responsibility related activities for the development of finger millet as a sustainable and nutritive food crop of the future. Similarly, enterprises engaged in the production of food products for baby, child and pregnant and lactating mothers may produce finger millet based food solutions for high nutrition and better health.

Acknowledgement: A draft version of this paper 'Climate Change and Sustainable Agriculture: Finger Millet for Food, Nutrition and Health Security' was presented in an International Conference on Challenges for Local Governance and Development in the 21st Century, Organized by International Association of Schools and Institute of Administration, International Institute of Administrative Sciences, Brussels, Belgium and National Institute of Development Administration, Bangkok, Thailand, July 16-21, 2012. The author is thankful to participants for discussions and suggestions. Usual disclaimer applies.

REFERENCES

Bala Ravi, S., (ND) "Neglected Millets that Save the Poor from Starvation", www.underutilized-species.org/documents/Publications/millet_mssrf.pdf, downloaded April 20, 2012, pp. 1-8;

Budget (2011-12) Union Budget, Government of India, Finance Minister, Budget Speech, in PDF Format http://exim.indiamart.com/budget-2011-12/budget-speech-2011-12.html downloaded 24.4.12.

Commission for Agricultural Costs and Prices (CACP) (2010) Reports of the CACP for the Crops Sown During 2009-10 Season, Department of Agriculture and Cooperation, Ministry of Agriculture, New Delhi, Government of India,.

Frison, Emile A.; Jeremy Cherfas and Toby Hodgkin (2011). "Agricultural Biodiversity is Essential for a Sustainable Improvement in Food and Nutrition Security", *Sustainability*, 3, *238-253.*

Global Facilitation Unit (GFU) for Under Utilized Species, Rome Italy, in http://www.cropsforthefuture.org/publication/Species-brochure/Species%20brochure-minor%20millets.pdf, downloaded, April 24, 2010.

Goplan, G.; B. B. Ramshastri and S. C. Balasubrahmaniam (1999). "Nutritive Value of Indian Foods", National Institute of Nutrition, Hyderabad, Cited in Sarita Srivastava (2008). *Madua Ke Paushtik Byanjan.* State Institute of Agriculture Management and Training, Uttarakhand, Extension Education Directorate, Pant Nagar, G.B. Pant University of Agriculture and Technology

ICRISAT (2007). "Group and Plenary Discussions and Recommendations" in M.A. Mgonja, J.M. Lenné, E. Manyasa and S. Sreenivasaprasad (Eds). *Finger Millet Blast Management in East Africa: Creating Opportunities for Improving Production and Utilization of Finger Millet,* Proceedings of the First International Finger Millet Stakeholder Workshop, Projects R8030 & R8445 UK Department for International Development–Crop Protection Programme, September 13-14, 2005 at Nairobi, ICRISAT, Kenya, SAARI, Uganda and Warwick, HRI, UK, pp. 158-174.

Kisandu, D.B.; W. Ntundu; W.Y. Marandu; and M.A. Mgonja (2007). "Germplasm Collection and Evaluation of Finger Millet in Tanzania: Challenges and Opportunities for Improved Production" in M.A. Mgonja, J.M. Lenné, E. Manyasa and S. Sreenivasaprasad (Eds.). *Finger Millet Blast Management in East Africa: Creating Opportunities for Improving Production and Utilization of Finger Millet,* ICRISAT, Proceedings of the First International Finger Millet Stakeholder Workshop, Projects R8030 & R8445 UK Department for International Development–Crop Protection Programme, September 13-14, 2005 at Nairobi, ICRISAT, Kenya, SAARI, Uganda and Warwick, HRI, UK, pp. 23-33.

Mgonja, M. A.; E. Manyasa; J. Kibuka; P. Kaloki; S. Nyaboke and G. Wandera (2007). "Finger Millet in East Africa: Importance, Blast Disease Management and Promotion of Identified Blast Resistant Varieties in Western and Nyanza Provinces of Kenya" in M.A. Mgonja, J.M. Lenné, E. Manyasa and S. Sreenivasaprasad (Eds.). *Finger Millet Blast Management in East Africa: Creating Opportunities for Improving Production and Utilization of Finger Millet,* ICRISAT, Proceedings of the First International Finger Millet Stakeholder Workshop, Projects R8030 & R8445 UK Department for International Development–Crop Protection Programme, September 13-14, 2005 at Nairobi, ICRISAT, Kenya, SAARI, Uganda and Warwick, HRI, UK, pp. 49-65.

Mackey, Desmond; Averil Brown; Roy Copeland and Louise Cooke (2007). "Evaluation of Finger Millet as a European Forage Crop: Plant Growth, Weed Control, and Seed Pelleting" in M.A. Mgonja, J.M. Lenné, E. Manyasa and S. Sreenivasaprasad (Eds.). *Finger Millet Blast Management in East Africa: Creating Opportunities for Improving Production and Utilization of Finger Millet,* ICRISAT, Proceedings of the First International Finger Millet Stakeholder Workshop, Projects R8030 & R8445 UK Department for International Development–Crop Protection Programme, September 13-14, 2005 at Nairobi, ICRISAT, Kenya, SAARI, Uganda and Warwick, HRI, UK, pp. 85-89.

Millet Network of India (MINI), Deccan Development Society (DDS), Food First Information and Action Network (FIAN), "Millets: Future of Foods and Farming", pp. 1-16. http//www.swarah.org/shikshantar/millets.pdf downloaded April 28, 2012.

Ministry of Finance (2012). *Economic Survey 2011-12,* Government of India, Oxford University Press, New Delhi.

Mitaru, B. N. and S. M. Githiri (2007). "Priorities for Millet Research in East and Central Africa" in M.A. Mgonja, J.M. Lenné, E. Manyasa and S. Sreenivasaprasad (Eds.). *Finger Millet Blast Management in East Africa: Creating Opportunities for Improving Production and Utilization of Finger Millet,* Proceedings of the First International Finger Millet Stakeholder Workshop, Projects R8030 & R8445 UK Department for International Development—Crop Protection Programme, September 13-14, 2005 at Nairobi, ICRISAT, Kenya, SAARI, Uganda and Warwick, HRI, UK, pp. 158-174.

Nadkarni, M. V. (1986). "'Backward' Crops in Indian Agriculture: Economy of Coarse Cereals and Pulses" Review of Agriculture, *Economic and Political Weekly*, 21 (38 & 39). A113-A115 and A117-A118, September 20-27

National Commission for Enterprises in the Unorganized Sector (NCEUS) (2008). "A Special Programme for Marginal and Small Farmers", Report, New Delhi, Government of India, December.

Oduori, Chrispus and B. Kanyenji (2007). "Finger Millet in Kenya: Importance, Advances in R&D, Challenges and Opportunities for Improved Production and Profitability" in M.A. Mgonja, J.M. Lenné, E. Manyasa and S. Sreenivasaprasad (Eds.). *Finger Millet Blast Management in East Africa: Creating Opportunities for Improving Production and Utilization of Finger Millet,* ICRISAT, Proceedings of the First International Finger Millet Stakeholder Workshop, Projects R8030 & R8445 UK Department for International Development–Crop Protection Programme, September 13-14, 2005 at Nairobi, ICRISAT, Kenya, SAARI, Uganda and Warwick, HRI, UK, pp. 10-22.

Okwadi, Julius (2007) "Importance and Characteristics of Finger Millet Processing in Uganda" in M.A. Mgonja, J.M. Lenné, E. Manyasa and S. Sreenivasaprasad (Eds.). *Finger Millet Blast Management in East Africa: Creating Opportunities for Improving Production and Utilization of Finger Millet,* ICRISAT, Proceedings of the First International Finger Millet Stakeholder Workshop, Projects R8030 & R8445 UK Department for International Development–Crop Protection Programme, September 13-14, 2005 at Nairobi, ICRISAT, Kenya, SAARI, Uganda and Warwick, HRI, UK, pp. 102-111.

Planning Commission (2011), "Faster, Sustainable and More Inclusive Growth: An Approach to the 12th Five Year Plan", Draft, Government of India, New Delhi, August.

Shanthakumar, G.; Bhag Mal; S. Padulosi and S. Bala Ravi (2010) "Participatory Varietal Selection: A Case Study on Small Millets in Karnataka". *Indian J. Plant Genet. Resources*, 23(1): 117-121

Shiferaw, B., J. Okello and Ratna Reddy (2009). "Challenges of Adoption and Adaptation of Land Water Management Options in Smallholder Agriculture: Synthesis of Lessons and Experiences" in Suhas P. Wani, Johan Rockström and Theib Oweis (Eds.), *Rainfed Agriculture: Unlocking the Potential*, Oxfordshire, UK, CAB International, pp. 258-275.

Srivastava, Sarita (Ed.). (2003). *Mote Anaj Madhumeh Rogiyoun Ke Liye Labhdayak.* Department of Food and Nutrition, College of Home Science, Pant Nagar, G.B. Pant University of Agriculture and Technology.

Srivastava, Sarita (2008). *Madua Ke Paushtik Byanjan.* State Institute of Agriculture Management and Training, Uttarakhand, Extension Education Directorate, Pant Nagar, G.B. Pant University of Agriculture and Technology.

Upadhyaya, H. D.; R. P. S. Pundir and C. L. L. Gowda (2007). "Finger Genetic Resources Diversity of Finger Millet — A Global Perspective" in M.A. Mgonja, J.M. Lenné, E. Manyasa and S. Sreenivasaprasad (Eds.), *Finger Millet Blast Management in East Africa: Creating Opportunities for Improving Production and Utilization of Finger Millet,* ICRISAT, Proceedings of the First International Finger Millet Stakeholder Workshop, Projects R8030 & R8445 UK Department for International Development–Crop Protection Programme, September 13-14, 2005 at Nairobi, ICRISAT, Kenya, SAARI, Uganda and Warwick, HRI, UK, pp. 90-101.

Vivekanand Hill Agriculture Research Institute (2007). *Mandua (Ragi) Kee Vaigyanik Kheti,* ICAR, Almora.

Wani, S. P.; T. K. Sreedevi; J. Rockström and Y. S. Ramakrishna (2009). "Rainfed Agriculture – Past Trends and Future Prospects" in Suhas P. Wani, Johan Rockström and Theib Oweis (Eds.), *Rainfed Agriculture: Unlocking the Potential*, Oxfordshire, UK, CAB International, pp. 1-35.

Wanyera, N. M. W. (2007) "Finger Millet (*Eleusine coracana*) (L.) Gaertn" in Uganda in M.A. Mgonja, J.M. Lenné, E. Manyasa and S. Sreenivasaprasad (Eds.), *Finger Millet Blast Management in East Africa: Creating Opportunities for Improving Production and Utilization of Finger Millet,* ICRISAT, Proceedings of the First International Finger Millet Stakeholder Workshop, Projects R8030 & R8445 UK Department for International Development–Crop Protection Programme, September 13-14, 2005 at Nairobi, ICRISAT, Kenya, SAARI, Uganda and Warwick, HRI, UK, pp. 1-9.

Worldwatch Institute (2009). *State of the World 2009: Confronting Climate Change,* London Earthscan.

13

Agrarian Society, Economic Reforms and the Concerns for Marginals: Opportunities and Challenges from Rural India

Ashish Saxena

Introduction

The main source of productivity in the informational economy is the accumulation of knowledge. However, 'the transition or shift from the launching of the First Five-Year Plan to the critical terminal years of the Second Plan was the opening. However, 'the transition or shift from industrial to informational economy is not the historical equivalent of the transition from agricultural to industrial economies and cannot be equated to the emergence of service economy' (Castells, 2001:100). The structural change in the information economy is that it transforms the labour market and generates demands for specific skills worldwide. The character of labour undergoes a shift from manual labour to intellectual labour. In the information economy, economic growth would depend on 'human capital' and less dependent on physical capital, which was dominant in the industrial economy. There is increasing importance of highly skilled labour in the information economy (Stehr, 1999).

In alignment with the global forces, inclusive development involves complex processes and procedures of institutional changes and the redistribution of economic and political power for attaining egalitarian society. Looking at the development landscape of India, it is being transformed from the initial phase of infrastructural development to the present age of liberalization, privatization and globalization. In social-structural terms, globalization is a historical process of transition from the agrarian-industrial, post-industrial and finally the stage of the information society (Y. Singh, 2000). It is to be emphasized that Indian society is highly diversified and premised on hierarchical principle resulting into unequal citizen in general and also regional inequality and rural-urban divide in particular. On the implications of globalization on developing countries like India, Kurian (1997: 122) rightly emphasized that it is not a remote experience or removed from the lives of the ordinary people of India. The local manufacturers and workers may be replaced. This is happening to the handloom weavers in different parts of India. Kurian also mentions that globalization usually benefits people who can play the game, but the chances are that the many who are resource less, uneducated, and making a living through traditional production activities, will come under various kinds of pressures. At this juncture, there is a need for the peasant[1] and family farm based food sovereignty model, which, prioritizes local production of food for local and national markets, negates dumping, and uses sustainable production practices based on local knowledge. Evidence shows that this model is potentially more productive per unit area, more environmentally sound, and far more capable of providing rural families with a decent life with dignity, while providing rural and urban consumers with healthy, affordable and locally produced food. However, the dominant, neo liberal agro export model is pushing peasant and family farm agriculture towards extinction.

One finds the crisis of development itself, howsoever well implemented the policy may be. The ethical questions asked were: development at whose cost and at what cost, and what constituted development itself. People's movements resisted increasing commoditization and monopolization of natural resources like land, water and forest, their unsustainable use and unequal distribution, exploitative power relations, the centralization of decision-making and disempowerment of communities caused by the development process. They asserted people's rights over natural resources and decision-making processes.

Movements of landless, unorganized labour in rural and urban areas, *adivasis, dalits,* displaced people, peasants, urban poor, small entrepreneurs and unemployed youth took up the issues of livelihood, opportunities, dignity and development. Moreover, the process of development itself resulted in large-scale displacement, destitution, centralization and destruction of resources. As rightly pointed out by

Sanjay Sangvai (2007:112) these processes and their ideology have intrinsically been against both the 'modernist' baggage of development and at the same time also against post-modernist tendencies to postulate each struggle as fragmented from another and the predominance of the 'politics of the local'. The struggles are different, issue based and localized, yet their context and purpose is overarching and interrelated.

With the 'new economic policy' and concomitant liberalization, privatization and globalization (LPG) policies, various existing organizations and new forums like *Azadi Bachao Andolan* have launched a campaign against the onslaught of national and multinational capital. The united struggle by local communities, trade unions and various organizations under the National Alliance of People's Movements (NAPM) against the infamous Enron Company in Konkan and more recently against special economic zones (SEZ) are examples of major struggles against LPG. The struggle against the World Trade Organization (WTO) and free trade by thousands of peasants in Karnataka under the banner of the Karnataka *Rajya Raitha Sangha* (KRRS), or against the World Bank-sponsored forestry project by Madhya Pradesh organizations are some more illustrative examples. The struggles against Enron in the mid-1990s and against the Coca-Cola Company in early 2000 alsc brought in sharp focus the collusion between the powerful political elite in India with MNCs (Sangvai, 2007).

In addition to the elitist development, globalization of the Indian economy has dramatically influenced social life in the country and the countryside, especially the so called marginal's (SC's/ST's and OBC's) of rural India. The state has mounted policies of economic liberalization, claiming that the policies will ultimately benefit the entire population but a number of social scientists have argued that marginalized groups will be adversely affected. In particular, they have emphasized problems such as uneven regional development, wages disparities and increased level of inequality. Even the large sectors of the 'backward classes and castes remain extremely hard-hit by inflation, contradiction insecure rural employment, the virtual impossibility of entry into lucrative export markets, continual poor levels of literacy, health and living standards, and overall social and political marginality. It is realized that in the present era of privatization, the State's indirect withdrawal from 'welfare' of service sector people, particularly *Dalits,* became vulnerable. Further the shift of decision making power into the hands of trans-national institutions like the World Trade Organization, International Monetary Fund, World Banks and others has severely reduced the sovereignty of national government and resulted in a very serious drift. In this regard we find that the *Dalits* are at the lowest end of the economic and social ladder and are quite obviously at the receiving end of the elitist culture promoted by urban world and by globalization.

Adding further to the discourse, Shiva (1998) too rightly said that "the present process of globalization threatens to undermine aspects of agricultural policy. The Trans-National Corporations (TNC) are not guided by any philosophy of welfare of the marginalized. Liberalization emphasizes competition, efficiency and providing commodity for the market. Such an approach goes against the prevalent practice of poor farmers who produce for their own use and their methods are based on traditional wisdom. Thus it appears that for majority of '*Dalit*' and poor, globalization has meant further deprivation. The agricultural census data as well as various micro level studies conducted in different regions indicate that small and marginal farms adopt diversified farming, despite very limited space and low sizes of holdings. Further it is the affluent categories of farmers who are going for it and not the *Dalits* in majority. Thus we have to improve the structure of the society to attain the fruits of globalization.

Liberalization, Globalization and Nation-State

Before going in details on the issue, one has to concentrate on the following premises:

1. Subsistence-agriculture is no longer the backbone of rural economies. Although it still has an important role in shaping rural landscapes in developing countries like India, its weight in rural economies *vis-à-vis* commercial economy is often low and declining.
2. Regarding rural India and cultivators it is relevant to quote Shanin (1986: 12) the only way to handle effectively contemporary social reality is through models and theories in which peasant family farms do not operate separately and where peasant economy does not merely accompany other economic forms but is inserted into and usually subsumed under a dominant political economy, different in type. Also peasant economies are being transformed (and even re-established) mostly by 'external' intervention, especially by the state and the multinational companies.....To emphasize further, 'rural society and rural

problems are inexplicable any longer in their own terms and must be understood in terms of labour and capital flows which are broader than agriculture . . . (where peasant economy) is inserted into and subsumed under a dominant political economy, different in type' (Shanin, 1986: 24).

3. Input-driven technologies, part and parcel of Green Revolution, have enhanced productivity but also have contrived concern for sustainability of production and natural resources, have clicked the global alarm.
4. The research and extension systems, in their efforts to generate, formulate and transfer new technologies are being guided by the *homo technicus* image of the farmers which excludes other quintessential components that account for farmers' rationale and behaviour. Farmers' problems are not merely technical, but cover wider domain of socio-cultural, economic and managerial considerations.
5. In case of small farmers falling into the marginalized communities, farming is livelihood and not a mere profession, therefore, the sociology, psychology and micro-political economy of this way of life influence the farmers' adoption behaviour towards improved agro-technologies.
6. Unfortunately, farmers' participation in research and extension programmes is highly inadequate despite their potential for making contributions (indigenous knowledge).
7. Fortunately, in the era of liberalization and globalization, the move away from equality of outcome towards equality of opportunity has gone together with a growing policy focus on investment in human capital. Keeping in view the so called excluded categories especially the rural scheduled castes and tribes, such strategies worked out at the level of policy, have quiet clear affinities with neo-weberian theory which sees people's capacities, including their skill and credentials, as determining their life chances in market societies. The answer for socialist government, then is, to help people to develop their capacities, social and economic outcomes might be shaped by the development of opportunities, rather than through direct interventions at the level of 'who gets what'. The distinction between equality of opportunity and equality of outcomes is, therefore, a distinction between enabling capacities and engineering consequences.

8. Poverty is defined not solely on the basis of income but in terms of a lack of basic capacities or capabilities – health, education and literacy, reasonable housing condition, safety – which allows individuals to participate in social membership and to make choice in respect of their own lives (Sen, 1999). The relation between poverty and social exclusion refers to a broader sense of being shut out from full social and economic participation. At this juncture it is to be emphasized that the concerning issue severely pertains to the SC's and ST's of rural India.

India has sought rapid economic growth through 'planned development'. This has entailed large-scale public investments involving land acquisition. Unfortunately, large number of people, especially marginalized sections, have been displaced from their original habitats to make way for these development projects. Such projects have also permanently changed the patterns of land-use and natural resources that previously prevailed in these areas. Often, such acquisition of land leads to displacement of people, depriving them of their livelihood and shelter, restricting access to their traditional resource base, and uprooting them from their socio-cultural environment. These have traumatic, psychological and socio-cultural consequences for the affected population, especially 'subalterns'. This thus results into doubly marginalization, i.e., dislocation of indigenous people from their native habitats and also poor compensation in the form of resettlement and rehabilitation. Generally, they become passive owing to their underdog position, however, in many instances they show resistance and adopt violent protests. Glaring exemplifications of contemporary resistances are the resistance of the Singur farmers against the West Bengal government, protests by villagers of Bhatta and Parsaul villages against compensation for acquired land and liberal land policy for the Yamuna Expressway in U.P. It is in this context that the amendment Bill of the Land Acquisition Act, 1894 gains credence.

As rightly pointed out by Ramesh Chand, P.A. Lakshmi Prasanna and Aruna Singh (2011: 26-27) that between 1970-71 and 2005-06, the total number of operational holdings in India increased from 71.01 million to 128.89 million and operational holdings area declined from 162.18 million ha to 156.62 million ha. This resulted in the reduction of average farm size from 2.28 ha to 1.21 ha. In the same period, the share of small and marginal holdings in operated area doubled. Smallholders now cultivate 42 per cent of operated land and constitute 83 per cent of total landholdings.

To substantiate further, some social scientists have considered the implications of economic liberalization and forecasted that marginalized groups will be adversely affected (Acharya, 1993). In particular, they have argued that economic liberalization and marketization will reinforce asymmetrical gender relations. Yet, these are predictions, based on experiences of the poor elsewhere in the developing world, rather than actual case studies in India. The vulnerability of labour has increased considerably under globalization because the volatility of the global market does not ensure stable jobs to workers and increased competition in the domestic labour market results in closures, downsizing and restructuring of units. As the International Labour Organization (ILO, 2004) in its report on 'Economic Security for a Better World' has pointed out, society at present in developing countries is 'a risk society' because most groups, particularly those at the bottom, are exposed to uncertainty and risks that emanate from new policy regimes. Their insecurity is multidimensional related to employment, income, work, job, and skills.

In short, sociological discussions of economic liberalization in India focus on class polarization in general terms, rather than on empirical evidence. For example, it is argued that: "... market forces will polarize classes in both agricultural and industrial sectors. The growing global demand for vegetables, fruits and flowers as Indian agriculture links up with international agri-business and the consequent shrinkage of the area under food crops is inimical to the interests of the poor who will be made to pay higher prices for their daily food requirements. The poor will become poorer and whereas the rich farmers will attain prosperity because of their linkages to the global market" (Panini; 1995:54).

In the Indian context there has been some documentation of the growing inequalities due to the implementation of liberalization policies. According to the National Council of Applied Economic Research, for example, wealthier households have grown and benefited from liberalization at a more rapid rate when compared to poorer households (Natarajan, 1999). More specifically, this research shows that while there is a smaller proportion of poorer households in India now than before economic reforms in 1991, richer households are reaping a far greater monetary reward over the corresponding period, with their household income rising dramatically. Such discourses which are replete with references to the expansion of the middle class (Kulkarni, 1995:88-94).

Some farmers have actively supported liberalization since it promotes labour-intensive sectors of the economy and contributes to rural development (Omvedt, 1993: 2708-10). By removing both price controls and restrictions on trade, the economic 'reforms' have created incentives to increase agricultural productivity, which in turn has increased employment opportunities. Yet, while certain sectors of the rural economy have undoubtedly prospered, large sectors of the 'backward' classes and castes remain extremely hard-hit by inflation, contraction insecure rural employment, the virtual impossibility of entry into lucrative export markets, continual poor levels of literacy, health and living standards, and overall social and political marginality (Ganguly-Scrase, 1997).

It is observed that the nation-state, through the adoption of an innovative constitution, the launching of Five-Year Plans, decentralization of polity and similar measures, unleashed forces of social transformation. The nation-state became the core institution and development with justice became the new goals. Although equality became the central value its translation into reality remained a great challenge. The globalizing phase initiated by the liberalization of the economy. The distinguishing features of this phase are: minimizing of the role of the State in economic development, enlarged role of civil society in social transformation and centrality of the market in the economy. As the market became the core institutional order, the consumer assumed central identity sidelining citizenship rights against the rapacity of the market. The core institutional order of the global age being the market, both capital and consumer got privileged.

We may thus say that the globalization of the Indian economy has dramatically influenced social life in the country. While the state has mounted policies of economic liberalization, claiming that the policies will ultimately benefit the entire population, a number of social scientists have argued that marginalized groups will be adversely affected. In particular, they have emphasized problems such as uneven regional development, wage disparities and increased levels of inequality. Unquestionably, economic liberalization has not benefited the very poor. Yet, what of the lower income segment of salaried workers? Patents were instated for living organism, impoverishment of human society ethically, ecologically and economically, though they bring commercial gains to a handful of corporations (Shiva, 1991).

In July 1991, when the first comprehensive statements of policy on economic and trade liberalization were formulated for India, it was under considerable international pressure to follow the policy prescriptions of the major international lending institutions including the IMF and The World Bank which advocated trade liberalization and export-led growth. The Indian government's structural adjustment policies have proposed reductions in public sector employment, limitations on agricultural subsidies, de-nationalization of banks and insurance companies and reductions in public expenditure that are envisioned to increased global competition.

In the early years of independence, agriculture had a central position in the national consciousness at the same time that it became inscribed within a technocratic, state-centred developmentalism. Anti-colonial nationalists prior to independence had been divided in their opinions about the relative importance of industry *vis-à-vis* agriculture in the context of a vast agricultural society like that of India. Some argued that modernization could occur only through industrialization, which was a structural precondition for the improvement of agriculture. Others maintained that the industrialization of agriculture through the widespread application of scientific knowledge was more pressing; assuming that industry and the rest of the economy would depend fundamentally on the mass production of agricultural output. The first decade of planning (1951-61) was known as phase of agricultural policies with several distinctive features. First, the policies during this phase bear a clear imprint of the *Nehruvian* ideology of industrialization of India and modernization of rural society. As Radhakrishna (1993) observes, "in the first period... the main thrust was on institutional and agrarian reforms as well as expansion of the agrarian base"

A technocratic strategy dubbed the 'green revolution' was introduced in the Indian countryside in the mid-1960s. Since then, its impact on the political economy of rural India has been the subject of enduring scholarly debate. Some analysts have claimed that the effects of the 'green revolution' have been positive in terms of promoting agricultural development. Others have argued the opposite, claiming that its spread has significantly undermined long-term agricultural sustainability, exacerbated rural poverty and hunger (despite growing national food-stocks), and intensified class and caste conflict in the countryside (Bardhan, 1985; Lipton, 1989; Shiva, 1989). A survey of the literature reveals that the technocratic strategy was 'top-down', a clear case of 'development from above'. For example, according to Frankel (1978), orientation of the agricultural strategy represented a monumental 'attack' on the principles of *Nehruvian* socialism and a 'retreat from the social goals of planning'. Hence, although much of the 'green revolution' technology was relatively scale-neutral in its application, its availability was (and is) generally restricted by inequitable distribution of land, income, and credit. In fact, the profitability of commercial agriculture has released an entrepreneurial dynamism among the wealthy land-owners.

Green Revolution to L.P.G.: Agrarian Shifts in Rural India

The land-use pattern of any region/country at any point of time is governed by the prevailing physical, economic and institutional-framework. The interaction of these parametres keeps the land-use system dynamic in the long term, but stable in the short term. In other words, the land-use pattern in India keeps evolving as the result of action and interaction of various factors such as physical characteristics (productivity, location, geography, etc.) of land, the institutional framework (social, legal, etc.) and the availability of resources (water, capital, labour, etc.). As the economy is growing, the demand for land for purposes other than agriculture is continuously increasing. The per capita availability of agricultural land in India has decreased from 0.46 ha in 1951 to 0.11 ha in 2001 with the likelihood of a further decline in the near future (Chaturvedi, Patil and Goswami, 2011).

It is a delicate balancing between the conflicting interests of preserving and maintaining the natural resources while utilizing them for development. The people who are causing land degradation and at the same time are the ones most affected by its consequences are the poor and marginal farmers, cultivating marginal lands. It is also true that they have none or very limited access to the decision-making process, no political power and most of their actions are not based on greed but on the need to survive.

Green Revolution technologies placed in the hands of the state reproduced the vision of the state as social engineer, but in a manner that placed the countryside at the centre of that engineering. In this context, the development of HYVs within national laboratories was hit upon as a solution out of dependency. Yet HYVs were efficient only in areas already well endowed with water, and with low person-to-land ratios that would lend itself to intensive, widespread commercial farming. This exacerbated already extant inequalities between rural areas as well as inequalities within the farming population. Those farmers with resources that were capable of taking advantage of widespread chemical farming were placed in an advantageous position from the outset. Consequently, at the same time that the Green Revolution underscored the role of agriculture within nationalist ideology, it also produced a system of large-scale commercial farming that gave rise to a class of surplus-producing farmers.

This history of the process by which large-scale, state-sponsored agricultural production rose to occupy a central place in the making of the modern Indian nation-state suggests that sovereignty, the driving political imperative of the post-colonial nation, is inevitably linked with the issue of food security. Talking about agrarian economy, we find that the agricultural growth during 1990-93 to 2003-06 reflects the impact of economic reforms on agriculture performance. The most important feature of this period is that agricultural growth decelerated sharply at the all-India level and in all regions. At the all-India level, the output growth decelerated to 1.74 per cent per annum during 1990-93 to 2003-06 compared with a growth rate of 3.37 per cent per annum during 1980-83 to 1990-93. At the regional level, during the same period, the growth rate of agricultural output decelerated highest from 3.55 per cent to 1.58 per cent per annum in the north-western region. The main reason for the deceleration of growth during the post-reform period was a visible deceleration in investment in irrigation and other rural infrastructure (Bhalla and Singh, 2009).

In India, there are competing demands on area available for cultivation from increase in rural habitation, forestation, urbanization and industrialization. Consequently, net sown area in the country has registered a rapid deceleration in its growth over time. During 1962-65 to 1980-83, net area sown rose at a rate of 0.15 per cent per annum at the all-India level – its growth rate decelerated to 0.11 per cent per annum during 1980-83 to 1990-93 and further to –0.05 per cent per annum during 1990-93 to 2003-06. Even, the economic reforms initiated during the early 1990s were expected to hasten the process of crop diversification from low value foodgrains to high value non-foodgrain crops. However, during the post-reform period, the yield growth rates of most of the important crops including wheat and rice, oilseed, sugarcane decelerated considerably compared with the pre-reform period 1980-83 to 1990-93. One may say that economic reforms and trade liberalization have failed to hasten the process of diversification in agriculture.

V. Shiva (1991: 9) suggests that the 'militant cultures' of nationalism that have emerged in the last fifteen years as essential moments of public identification in India are a reaction to the homogenizing processes of global capital, on the one hand, and to a perceived decline in the power of the Indian State on the other. Conversely, the ethnicized territory of peasant nationalism is – unambivalently – a site of collective resistance against the 'West's paradigm of development' and the homogenizing cultural forces of a capitalism represented by a few multinationals and the superpowers that support them. Thus the social dislocations induced by development (as when populations are moved to make way for dams, roadways, plants, and so on) are construed as 'uprooting' that displace 'the rhythms and patterns of the universe.' Shiva (1989) is right to draw attention the dislocations of local communities. Indeed, it is precisely such dislocations that need to be highlighted as sites of injustice and as sites of resistance against national and transnational governmentalitiy.

Development is associated with the death of the 'sacred'. '"Development has meant the ecological and cultural rupture of bonds with nature, and within society, it has meant the transformation of organic communities into groups of uprooted and alienated individuals searching for abstract identities" ..."In effect, the process of development leads to turning away from the soil as a source of meaning and survival, and turning to the state and its resources for both. The destruction of organic links with the soil also leads to the destruction of organic links within society" (Shiva, 1991: 99). GATT and its enforcement through the WTO establish a neocolonial platform through which capitalist dominates nature. The new agents of agriculture – agribusinesses – have a very different relationship to the earth than the romanticized indigenous farmer. Rather than sustainably and harmoniously cultivating the soil, agribusiness is seen as a predator that tends to exploit and exhaust it: on Shiva's account, in addition to affecting the freedom of small farmers to steward the land, free trade's preference for corporate agriculture degrades the environment. "Local control over natural resources is an essential precondition for farmers' freedom. But free trade which, as we have seen, implies a relocation of control over natural resources for farmers and Third World governments to global institutions has serious environmental consequences" (Shiva, 1991: 233).

Multiple Crises: Food, Water, Livelihood, Man

No doubt, the green revolution technologies have helped in the reduction of poverty but have resulted in widespread regional disparities as dry lands are not covered. It took a long time for the technology to spread to these areas and even then rainfed farmers did not benefit to the same extent as those in well-endowed regions. This is because of the exclusive emphasis on increasing production without keeping the objective of equity in view.

The green revolution model, and now the globalization model of agriculture are based on the destruction of diversity, uprooting of small peasants through indebtedness, chemical intensification, capital intensification, centralization and long distance transport. They are based on intensification of external inputs, and integration with global markets. In the globalization era, the commons are being enclosed and the power of communities is being undermined by a corporate enclosure in which life itself is being transformed into the private property of corporations.

Today we have to look beyond the State and the market place to protect the right of the majority of Indians – the rural communities. The 'commons' cannot be monopolized by the economically powerful citizens or corporations or by the politically powerful State. While the tribal and rural communities are still overwhelmed by State-driven 'enclosures', tools for new corporate and WTO-driven 'enclosures' are being shaped in the form of patents on life and biodiversity.

Trade liberalization is leading to external liberalization, or liberalization for global agribusiness. What the Indian farmers need is internal liberalization. For example — the promotion of bio-fuel development is attractive for a country like India because of its potential for creating employment opportunities for the rural poor as well as offering opportunities for promoting local level entrepreneurship and enhancement of women's participation. The availability of technologies for decentralized production of bio-fuels offers opportunities for the development of local level entrepreneurship. Local institutions like Joint Forest management committees, self-help groups and Panchayat can play an important role in involving village communities in bio-fuel programmes. Locally produced bio-fuels can provide fuel for irrigation pump sets and for electricity generation, which will improve access to modern energy services to rural population and help in improving productivity.

However, any large-scale bio-fuel programme has to ensure that it does not compromise with the nation's food and nutritional security. Further, utilizing unirrigated, barren and wasteland for growing non-food bio-fuel crops would prove fruitful with multiple benefits like greening up of unused lands, creation of employment opportunities, greater people's participation, and so on.

Environmental standards and regulations have been relaxed, or allowed to be ignored, in the bid to make the investment climate 'friendlier' to both domestic and foreign corporations. Governments are sacrificing more and more natural habitats and prime food-growing land to make way for commercial enterprises. The need for infrastructure or raw materials or commercial energy is determined not by the imperatives of human welfare and equity, but by economic growth rate targets, even where the growth rates may have no necessary co-relation with human welfare. The last couple of decades have, therefore, seen a massive increase in new infrastructure creation (highways, airports, urban infrastructure and power stations). This has meant increasing diversion of land, mostly natural ecosystem like forests or farms and pastures.

As an alarming situation, it is rightly pointed out by V. Shiva (2010:23) – Two aspects of the food and agrarian crisis in India are the epidemic of farmer's suicide and increase in hunger. Per capita calorie intake has fallen from 1830 Kcal in 1989 to 1600 Kcal in 1998. In 1999-2000 almost 77 per cent of the rural population consumed less than the poverty line calorie requirement of 2400 calories. Farmer's suicide and increase in malnutrition are related processes – both are the result of ten years of rapid, unthinking trade liberalization. Globalization has led to:

1. Increase in cost of production.
2. Falling prices of farm produce.
3. Rising costs of food as a result of the dismantling of the PDS system and weakening of the Essential Commodities Act.

Main impact of liberalization of industrial policy has been a shift from developmentalism based on the strategy of growth with justice towards an elite-oriented model of growth. The green revolution model, and now the globalization model of agriculture are based on the destruction of diversity, uprooting of small peasants through indebtedness, chemical intensification, capital intensification, centralization and long distance transport. They are based on intensification of external inputs, and integration with global markets.

A very large section of India's population is going through severe and multiple crises: food insecurity, water shortages, inadequate fuel availability, and dislocation of livelihoods with limited alternative options. Finally, there is the crisis of livelihoods. As ecosystem disruption and land/water degradation intensifies, or as access to natural resources and traditional consumer declines, communities who have been traditionally self-employed (as farmers, pastoralist, fishermen, hunter-gatherers) are increasingly impacted.

The debate on agricultural biotechnology is focused mainly on the environmental impact, bio-safety issues, and intellectual property rights. The State will have to play a proactive role in priority setting for pro-poor agricultural development and accessing tools and technologies from the private sector by forging creative public-private partnerships to focus on crops and problems of importance to small farmers. It is to be stressed that ecologically, patent related to biological resources have major implications for the conservation of biodiversity and its sustainable use.

By creating 'property' in life through patents, the economically powerful corporations can become the new 'life-lords', just as earlier we had landlords or zamindars.

Globalization and Rural *Dalits* in India[2]

The poor remain poor as they are unable to utilize their land resource to the optimum. There is always a dearth of money for investing in fertilisers, pesticides and all the ingredients required for modern agriculture. And because there is no investment, productivity remains poor completing the cycle. There is no doubt that poor people with poor information and technologies can never succeed in improving their lot on poor lands. If this vicious circle is to be broken at least one variable has to change.

In the context of globalization and development, *Dalits* at the lowest end of the economic and social ladder are quite obviously at the receiving end of the elitist culture promoted by globalization. They are victims of globalization in both the economic and the social sense. The bulk of *Dalits* are without land and so land reform is, therefore, a key question for their development. But, globalization has not only eclipsed this agenda of land reforms it has substituted it with corporatisation of farming for the global agricultural market. The general population is facing the horrors of unemployment; this will be more deeply felt by *Dalits*. Here, it becomes crucial to observe – What are the implications of this paradigm change for those who are weak, poor, marginalized and downtrodden (like Scheduled Castes)? Can Nation-State tackle the "social concern" (welfare goals) for '*Dalits*' in the globalization era? To put it differently, what would be the implications of the globalization on eradication of social inequality and other forms of disparities (especially for '*Dalits*')? Would the process protect and promote human rights of the *Dalits*? In nutshell what are the major implications of globalization for *Dalits* in Indian context.

As per '*Peoples March Group*' analysis (2003) — more than 75 per cent of the *Dalit* workers are still connected with land; only 25 per cent of which are marginal and small farmers, while the balance is landless labourers. To Thorat (1996), the new economic policy also affected the *Dalits* adversely. After the implementation of the economic reforms in 1990s, both unemployment and the poverty levels in the country have increased. The number of *Dalit* employees in government service declined from 628,000 in 1991 to 369,000 in 1992.

Two questions, who you are? where you are? makes crucial differences and give very different grounds and possibilities of action. Globalization can affect the social space of actors (here for Scheduled Caste) from two angles: effect from the top, i.e., by directly changing their given social location in terms of consumerism, competition and development or effect from below, i.e., by opening to them the new avenues from rest of the world . Thus globalization being a central driving force behind the rapid social, political and economic changes can reshape the modern societies and world orders and no society escapes from its reach as it re-casts traditional patterns, creates new hierarchies, and most crucially reshapes the power, functioning, authority and objective of nation-state.

Taking another dimension, Shiva (1998) said that "the present process of globalization threatens to undermine aspects of agricultural policy. The trans-national corporations (TNC) are not guided by any philosophy of welfare of the marginalized. Liberalization emphasizes competition, efficiency and providing commodity for the market. Such an approach goes against the prevalent practice of poor farmers who produce for their own use and their methods are based on traditional wisdom. Land has been transferred to TNC's who produce luxury and non-food crops (cash crops) for export rather than staple food. There is also diversion of cereals from domestic market to export, resulting insecurity and inflation. The farmer's suffers from food insecurity as he has lost his land, resources and is unable to buy them from the market" The contract farming, which is promoted as a part of globalization had adverse effect on the farmers. The contract made with illiterate farmers is more one sided. Further the traditional societies emphasize cooperation; harmony and accommodation with personal relations as existed through *jajmani* and other caste based system. But the globalism emphasizes efficiency and competitive culture. Thus it appears that for majority of '*Dalit*' and poor, globalization has meant further deprivation.

The economic dimension of India can be seen through agriculture land and related issues. In Indian situation as the bulk of *Dalits* are without land so land reform is a key solution for their development. But, globalization has not only eclipsed this agenda of land reforms it has substituted it with corporatization of farming for the global agricultural market. The thrust of the World Bank-IMF-WTO combine has always been clearly for the abolition of the land ceiling laws and for liberalizing investment into agriculture. Land reform has openly been opposed. If we see the situation of *Dalits* in agriculture we find that they will not be immediate beneficiaries.

According to Prof. Haq (1998) diversification of small farms is often suggested as a means for rapid rural development in India. He emphasized on the need for commercialization and diversification of small farms within and outside agriculture and their proper integration with local and global markets. This is intended not only to liberate the small and marginal farmers from the poverty trap, but also to meet the country's growing demands for fruits, vegetables, milk and milk products, meat, fish, eggs, etc., which generally show rising trends with increasing levels of per capita income in the economy. However, the moot question today is how do we achieve the goal of diversification of small farms by overcoming various technological, infrastructural, institutional and policy constraints. The agricultural census data as well as various micro level studies conducted in different regions indicate that small and marginal farms adopt diversified farming, despite very limited space and low sizes of holdings. Further it is the affluent categories of farmers who are going for it and not the *Dalits* in majority. Thus we have to improve the structure of the society to attain the fruits of globalization.

This relationship cannot be altered till the *Dalits* get land. Thus, what impact would the new regime have on the socio-economic setting of Indian villages? Can the new regime afford this economic empowerment of the *Dalits*? Can it, for instance, grant them land? The answer to all these questions will be in negative. Instead of talking about land reforms, the new regime will promote depeasantisation of Indian agriculture and consolidation of their holdings to start corporate farming. Thus the privatization and free market components of the reforms are certainly affecting very adversely on the livelihoods of *Dalits*. The impact of the reforms already made the *Dalit* society more insecure and no doubt it will uproot the entire community if we allow it in the same fashion.

Concluding Remarks

The discourse of the chapter thus revolves around – implications under globalization for the rural India in general, and especially for those who are weak, poor, marginalized and downtrodden in the area. To put it differently, it reflects the level of inequitable inclusion of the rural population and the marginals. Gradual decline in the nation-state policy to tackle the "social concern" (welfare goals) for marginal peasantry is emphasized along with an impression that the new regime promotes more inequality, starvation, indebtedness, suicides, deprivation and depeasantisation *vis-à-vis* agriculture and consolidation of their holdings in favour of corporate farming. In brief:

1. Globalization has effected both social relations of production and social relations of consumption in rural India.
2. Urgent need in today's world is democratizing rural economy. In today's conservative, neoliberal political environment, strong grassroots poor people's movements is critical to pushing the reform process, stopping government foot dragging and, when necessary, taking matters into their own hands.
3. A new era is being constructed in which state, people, and the environment are forced to adjust to the market, and it is the markets, not people, that require freedom.

REFERENCES

Acharya, S. K (1995). 'Spatial Implications of the New Economic Policy — Reflections on Some Issues for North East India with Special Reference to Women', *Man and Development,* Vol. 17, No. 1, pp. 56-64.

Bardhan, P. (1985). 'Poverty and trickle-down in rural India: A quantitative analysis 'in J. Mellor and G.N. Desai (eds.): *Agricultural Change and Rural Poverty,* Baltimore: John Hopkins University Press.

Bhalla, G. S. and Gurmail Singh (2009). 'Economic Liberalisation and Indian Agriculture: A Statewise Analysis', December 26, Vol XLIV No 52 *Economic & Political Weekly*, pp. : 34-44.

Castells, M. (2001). *The Rise of the Network Society, The Information Age: Economy, Society and Culture*: Volume 1, Blackwell Publishers, Great Britain.

Chand, Ramesh; P. A. Lakshmi Prasanna and Aruna Singh (2011). 'Farm Size and Productivity: Understanding the Strengths of Small Holders and Improving Their Livelihoods'; *Economic & Political Weekly* Supplement *EPW*, June 25, Vol. XLVI, pp. 26-27.

Chaturvedi, Arun; N. G. Patil and S. N. Goswami (2011). 'Reorienting Land Use Strategies for Socio-economic Development in Uttar Pradesh', *Economic & Political Weekly* Supplement *EPW*, June 25, Vol. XLVI, pp. 26 & 27.

Frankel, F. (1978). *India's Political Economy 1947-77*, Princeton: Princeton University Press.

Ganguly, R-Scrase. (1997). 'We Want a Piece of the Action: Structural Adjustment and Indian Leather Workers'. Paper presented at the Asia

Pacific Regional Conference of Sociology *Social Transformation in Asia,* 18-21 Sept., University of Malaya, Kuala Lumpur.

Gupta, Akhil, (1998). *Postcolonial Developments: Agriculture in the Making of Modern India*, Durham, N.C.: Duke UP.

Haq, Mahbub-ul (1998). Quoted in *Foreword* by Khadija Haq in Report on *Human Development in South Asia 2001-Globalization and Human Development*. Oxford University Press: New York.

Kulkarni, V. G. (1993). 'The Middle Class Bulge', *Far East. Econ. Rev.*, Vol. 156, No. 2, pp. 44-46.

Kurian, C. T. (1997). "Globalization—What is it about?" In *Integral Liberation*, Vol. 1, No. 3, Sep., 122-40.

Lipton, M. (1989). '*New Seeds and Poor People*', Baltimore: John Hopkins University Press.

Natarajan, I. (1999). '*India Market Demographics Report,* National Council of Applied Economic Research', New Delhi.

Omvedt, G. (1993). 'Farmers Movement: Fighting for Liberalization', *EPW,* Vol. XXVIII, No. 50, pp. 2708-10.

Panini, M. (1995). 'The Social Logic of Economic Liberalization', *Sociological Bulletin,* Vol. 44, No. 1, p. 54.

Radhakrishna, R. (1993). 'Indian Agriculture: Growth, Employment and Equity' in Tapas Majumdar (ed.) *Nature, Man and the Indian Economy*, New Delhi: Oxford University Press.

Sangvai, Sanjay. (2007). 'New people's movement in India', 15 December, *EPW*; 112.

Sen, Amartya. (1999). *Development as Freedom,* New Delhi: Oxford University Press.

Shanin, T. (1986). Chayanov's message: illuminations, miscomprehensions, and the contemporary 'development theory'. In: A.V. Chayanov, *The Theory of Peasant Economy*, eds. D. Thorner, B. Kerblay and R.E.F. Smith, second edition. Madison, WI: University of Wisconsin Press, pp. 1–24.

Shanin, T. (eds.). (1971). *Peasant and Peasant Societies*, Harmondsworth, Penguin Publication

Shiva, V. (1989). *The Violence of the Green Revolution: Ecological Degradation and Political Conflict*, London: Zed Press

——. (1991). *The Violence of the Green Revolution: Third World Agriculture, Ecology, and Politics*, Penang, Malaysia: Third World Network.

——. (1995). *Captive Minds, Captive Lives: Ethics, Ecology, and Patents on Life*, Dehra Dun, India: Research Foundation for Science, Technology, and Natural Resource Policy.

——. (1998). *Globalization of Agricultural, Food Security and Sustainabilities*, Research Foundation for Science and Technology, New Delhi, June.

Shiva, V. (2010). The social costs of economic globalization, in P. J. George (eds.) *Reflections on Vital Issues*; Hyderabad: Orient Blackswan, p. 23.

Singh, Y. (2000). *Culture Change in India – Identity & Globalization.* Jaipur: Rawat Publication.

Stehr, N. (1999). 'The Productivity Paradox: ICTs, Knowledge and the Labour Market, in John de la Mothe and Gilles Paquet (Eds.), *Information, Innovation and Impacts*, Kluwer Academic Publishers: Boston MA.

14

Contemporary Debates on Food Security in South Asia

Sunil Kumar Tripathi

Introduction

Large-scale policy and plans have been implemented to remove the poverty and hunger but the battle against this is far from won. According to the latest report from the Food and Agriculture Organization (FAO) of the United Nations, 303 million people in South Asia are food insecure, i.e., they lack sustainable access to sufficient food to lead healthy and productive lives. Presently, about 40 per cent of the food-insecure people in the developing world live in South Asia. According to an estimate about 260 million people were hungry and deprived in year 2009-10 (FAO, 2010). 45 per cent of our children were malnourished and underweight and India ranks 126th out of 177 countries in the Human Development Index (UNDP, 2010). The national family health Survey (2008) showed that the child undernutrition rate is 46 per cent which is almost double that of sub-Saharan Africa. In the Global Hunger Index (2009), India ranks 65th among 84 countries (IFPRI, 2009) slightly above Bangladesh and below all other South Asian countries even India comes below Sudan, Nigeria and Cameroon.

Right to food is a birth right for all and should be an integral part of the right to life, as enshrined in Article 21 of the Constitution of India as well as Universal Declaration of the Human Rights (1948). Unless it gets enforced legally and socially, hunger will continue. The Government of

India announced in 2009 that a Food Security Act will be passed to ensure a particular quantum of wheat and rice to needy households. Since then there have been debate on various aspects of the food security act and its implications in money terms. The government is making efforts to address the issue of hunger and undernutrition in the country so that every citizen does not go to bed hungry while the country grows at one of the fastest pace in the world (Singh, 2011).

India currently possesses as on 1st June 2010 the stock of foodgrains in the central pool rose as high as 60 million tonnes whereas the prescribed stock limit in July is only 26.9 million tonnes. Paradoxically, though India is one of the largest producers of the food in the world, yet millions of people struggle to get two square meals a day and an equal number are undernourished (IFPRI, 2009). While efforts need to be made to increase the production and productivity of the foodgrains, providing access to the needy remains a major challenge.

Food Security: The Concept

As per the World Bank's definition, the food security is 'access by all people at all times to enough food for an active and healthy life.' However, two commonly used definitions of food security come from the UN's Food and Agriculture Organization (FAO) and the United States Department of Agriculture (USDA): Food security exists when all people, at all times, have physical and economic access to sufficient, safe and nutritious food to meet their dietary needs and food preferences for an active and healthy life. Thus common to most definitions of food and nutrition security are the elements of availability, access, utilization and stability or sustainability (Hahn, 2000). Given the stress on availability, access, utilization and stability in defining food security, it is important to understand these concepts.

Availability, a function of production and distribution, refers to the physical existence of food, be it from own production or on the markets.

Access, a function of purchasing power and employment, is ensured when all households and all individuals within those households have sufficient resources to obtain appropriate foods for a nutritious diet (Riely, *et al.*, 1995)

Utilization, a function of clean drinking water, sanitation and healthcare, has a socio-economic and a biological aspect. Utilization requires not only an adequate diet, but also a healthy physical environment, including safe drinking water and adequate sanitary facilities (so as to avoid disease) and an understanding of proper health care, food preparation, and storage processes.

Stability or **Sustainability** refers to the temporal dimension of nutrition security, i.e., the time frameover which food security is being considered. Generally two types of food insecurity is mentioned in food security review literature, chronic food insecurity—the inability to meet food needs on an ongoing basis—and transitory food insecurity when the inability to meet food needs is of a temporary nature (Maxwell and Frankenberger, 1992).

Food security has to be sustainable over time. And it is not just a matter of the quantity of food. The quality of food, that ensures nutrition, and healthcare that protects people from debilitating diseases or intestinal parasites that leak nutrition, are all part of it (Radhakrishna, 2005).

Food Security and the South Asian Scenario

Major changes have taken place in the food production scenario of South Asia over the last few decades. India, the largest of the South Asian economies, is now largely self-sufficient in foodgrain and an emerging exporter. While the other countries of the region remain dependent on cereal imports, the per capita availability of cereals has increased in every country (with the exception of the Maldives) from 1980s till about 2002 (FAO, 2002, cited in Ramachandran, 2008).

Yet, endemic pockets of hunger remain, seasonal shortfalls are manifest and malnutrition is widespread across the region, women and children being the greatest sufferers. The 'Asian Enigma' as it is termed, of food scarcity and malnutrition amidst plenty, has challenged all attempts at resolution. Poverty alleviation strategies, livelihood generation programmes and direct food interventions have all been tried but to little avail. Food security researchers have often commented on the fact that, while most South Asian countries have available food stocks and better health and education services in comparison to many other developing countries, even most countries of food deficit Sub-Saharan Africa, score higher in terms of the nutrition levels of their women and children (Ramachandran, 2008).

The Global Conference on Agricultural Research for Development (GCARD) background report (GCARD 2010, March 28-31, Montpellier, France), highlighting the importance of the Asia-Pacific region in the global agrarian economy noted that the region is "a major supplier of most of the important food and agricultural commodities, accounting for production and consumption of over 90 per cent of the world's rice, 70 per cent of vegetables, 60 per cent of seed cotton and 45 to 50 per cent of cereals, oil seeds and pulses. With 3.5 billion people, the region accounts for about 58 per cent of the world's population. Of these, 2.45 billion live in rural areas, 70 per cent of the region's humanity. Agriculture (crops, livestock, fishery, forestry, and the associated natural resources endowments and distribution systems) is a source of livelihood for nearly 2 billion, about 80 per cent of the rural people in the Asia-Pacific. Thus, the region accounts for 74 per cent of the world's agricultural population" (Singh, 2009: 2).

But, the Asia Pacific region has only 38 per cent of the world's agricultural land of which, fortunately 32 per cent is irrigated against only 10 per cent in the rest of the world, thus accounting for two-third of the world's irrigated area. Land availability per person in agriculture in the region (0.3 ha) is almost one-fifth of that in the rest of the world (1.4 ha), and over 80 per cent of the world's small and marginal farmers belong to this region. Thus the region's agrarian landscape is dotted essentially by smallholders, nearly 40 per cent of who are resource-poor, numbering about 500 million farm people – farmers and their families (Singh, 2009).

Agro-ecological, size, socio-economic, cultural and demographic heterogeneity is a defining characteristic of the region. South Asia still tops in the home of the poor in the world with about 400 million people living below poverty. The numbers of undernourished, under-weight and under-height children and of low birth-weight infants in the South Asian region are substantial. Agriculture is the main source of livelihood of people in these countries.

New opportunities are emerging in the form of demographic advantages (more young people), new technologies (biotechnology, nanotechnology, information technology), diet revolution and changing demands, emerging value chains and super markets and the entry of private sector. The recent rise in food prices of basic foods has sent a shock wave through the world community, particularly poor people, arousing individuals and institutions from years of complacency about the state of the agricultural sector. Investments in agricultural research typically rank first or second in terms of returns to growth and poverty reduction, along with investments in infrastructure and education. Fortunately, there is a consensus and also action in these countries towards higher investment in agriculture and related areas. The obvious questions in this context are how much this investment should be and where should it be focused (Singh, 2009). South Asia has witnessed high growth rate of GDP in the period 1993-2006. However, the growth rate of agriculture has been slower in the corresponding period (Figure 14.1).

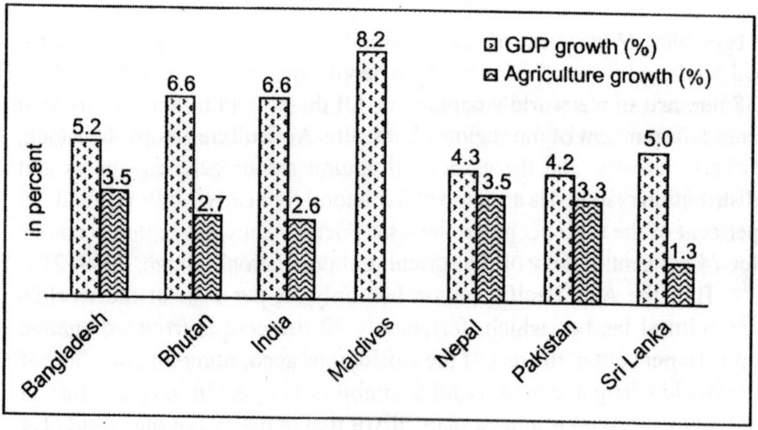

Figure 14.1: Average Annual Growth Rate (GDP and Agriculture) 1993-2006
Source: Mittal and Sethi, 2009, (using World Development Indicators, 2008).

In the same period, these countries witnessed a high annual growth rate of consumption that is often attributed to a high population growth rate Figure 14.2 (Mittal and Sethi, 2009).

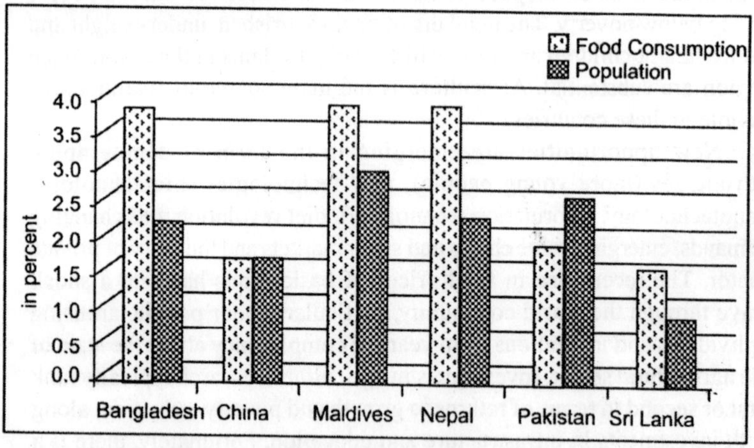

Figure 14.2: Average Annual Growth Rate of Food Consumption and Population, 1995-97 to 2001-03
Source: Mittal and Sethi, 2009, (Using FAO data, latest available figures of 2000-02).

Issues of Debate on Food Security

There are so many contemporary issues on which the discussion can be made but following are the most important which cannot be left over.

The Land Size Productivity Debate

An inverse relationship is said to exist between farm size and productivity (defined as output value per unit of cropped area), something that has puzzled economists for a long time. Chayanov (1926) first documented that small farms produce more output per cropped area in the Russian agriculture.

For India, the same evidence was found by Sen (1962), Bardhan (1973), and Rosenzweig and Binswanger (1993). The debate in the South Asian context was triggered off by Sen (1962) in which he stated by and large, productivity per acre decreased with increase in size of holding. The inverse relationship was derived based on land size class data; and Sen himself was, however, aware of the limitation of his conclusion since he was using only aggregated data. Sen (1964) subsequently gave three alternative lines of explanation for this phenomenon, (i) technique-based, (ii) labour-based, and (iii) fertility-based.

Later researchers have argued that the relationship becomes positive with the advent of green revolution. However, some researchers have declined to accept any kind of relationship between these two variables On the whole; authors on this issue may be categorized into two major groups. Those who have argued either a positive or negative relationship are Khusro (1964); Mazumdar (1975); Rao (1966); Bhagwati and Chakravarty (1969); Bhardwaj (1974); Bardhan (1973) and Hossain (1977). The other group of researchers who do not agree on the existence of a relationship (either inverse or positive) between farm size and land productivity are Rudra (1968) and Chattopadhyay and Rudra (1984). They have mainly regarded this relationship as statistically spurious for it does not explain the role of numerous factors governing the relationship between farm size and land productivity.

Mahesh (2000) examined whether the increase in productivity of land is explained by the decrease in the size of holdings. While the agricultural census figures suggest a possible relation between the two, the data collected through a sample survey in a rural locality in Kerala showed interesting results. A size-wise analysis of productivity indicated that the large farms had higher productivity. But a more detailed analysis, using regression techniques, failed to show a firm relationship between the size of the farm and its productivity.

To sum up, it is often pointed out that the difference in the size of farms is one of the reasons for the difference in yields. However, studies carried out on the relation between size of farms and productivity show contradicting results. Studies based on aggregated data showed an inverse relationship, but studies based on disaggregated data failed to confirm this. The latter indicates that the inverse relationship exists in certain types of farms, but the relation cannot be generalised. The debate thus remains inconclusive.

Falling Share of Agriculture in GDP

The falling share of agriculture in GDP has been not only a cause for concern, but evidence for the case against small-holder agriculture in Asia. In the Indian case, although its share in GDP has declined from over half at Independence to less than one-fifth currently, agriculture remains the predominant sector in terms of employment and livelihood with more than half of India's workforce engaged in it as the principal occupation. Agriculture still contributes significantly to export earnings and is an important source of raw materials as well as of demand for many industries.

India's agriculture sector has an impressive long-term record of taking the country out of serious food shortages despite rapid population increase. This was achieved through a favourable interplay of infrastructure, technology, extension, and policy support backed by strong political will. The main source of long-run growth was technological augmentation of yields per unit of cropped area. This resulted in tripling of foodgrain yields, and foodgrain production increased from 51 million tonnes in 1950-51 to 217 million tonnes in 2006–07. Production of oilseeds, sugarcane, and cotton have also increased more than four-fold over the period, reaching 24 million tonnes and 355 million tonnes and 23 million bales, respectively, in 2006-07.

Growth of agricultural GDP decelerated from over 3.5 per cent per year during 1981-82 and 1996-97 to only around 2 per cent during 1997-98 and 2004-05 (Table 14.1). Consequently, growth of agricultural GDP has been well below the target of 4 per cent set in both Ninth (1997-2002) and Tenth Plans. The recent trend towards diversion of food crops for bio fuels in surplus countries in South Asia means that food security needs a stronger production response based on tackling supply side problems in the foodgrains sector.

The supply side performance of agriculture is affected by a large number of factors, several of which interact among each other. These factors are the natural resource base (including rainfall), technology, infrastructure (including irrigation), and the economic environment comprising price signals and institutions

Table14.1: Average GDP Growth Rates – Overall and in Agriculture

(Per cent per Year at 1999–2000 Price)

Sl. No.	*Period*	*Total Economy*	*Agriculture and Allied Sector*	*Crops and Livestock*
1.	Pre-green revolution 1951-52 to 1967-68	3.69	2.54	2.65
2.	Green revolution period 1968-69 to 80-81	3.52	2.44	2.72
3.	Wider Tech. dissemination period 1981-82 to 90-91	5.40	3.52	3.65
4.	Early reform period 1991-92 to 2001-02	5.69	3.66	3.68
5.	Ninth Plan period 1997-98 to 2001-02	5.52	2.50	2.49
6.	Tenth Plan period 2002-03 to 2006-07	7.77	2.47	2.51
	Of which 2002-03 to 2004-05	6.60	0.89	0.89
	2005-06 to 06-07	9.51	4.84	4.96

Source: National Accounts Statistics 2008 (New Series), Central Statistical Organization, Ministry of Statistics and Programme Implementation, New Delhi.

Impact of Trade Liberalization and Related Reforms on South Asia

Since 1991 India (other countries too of South Asia around this period) initiated reforms as a part of the structural adjustment process, involving, among other things, change in exchange rate and liberalization of external trade. These structural adjustment programmes involved macro-economic contraction (lower public expenditure) and reduction in the developmental role of the State. The essential idea behind such programmes is that private investment will rise when public investment declines. However, the problem remains that specific direction of private investment will always be motivated by private profitability and will not involve consideration of infrastructure, employment generation or poverty alleviation. In India, measures have been taken to promote integration of domestic economy with global economy. These changes in turn have affected domestic prices of several commodities and, terms of trade for agriculture have undergone changes during the decade 1990s. Along with liberalization of trade, private sector has been allowed and encouraged to participate in the import and export of major agricultural products. Domestic prices of cereals, through government intervention in minimum support prices and open market operations, were given substantial hikes to reduce protection to agriculture because of domestic prices being lower than the international prices of several commodities. This way, prices became the major driving force for growth and development of agriculture sector during 1990s. The process of reforms, particularly trade reforms, further intensified since 1995 following implementation of WTO agreement on agriculture.

When reform period is divided into pre WTO reforms and post WTO reforms the growth rates present very depressing scenario – sharp deceleration and negative growth rates in some cases. Similarly, exports were adversely affected and imports have seen fast growth after WTO. There was a modest increase in per capita income of farm and labour households during reforms, contributed more by non-farm incomes than farm incomes. There was also a significant reduction in poverty. However, household food security and nutrition worsened during reforms, due to the high growth in prices of cereals caused due to government policy to give substantial hikes to cereal prices during reforms (Chand, 2004).

The Role of Subsidies

The role of subsidies to agriculture in South Asia remains contested. In India, subsidies are provided primarily for the following inputs: fertiliser, power, irrigation and credit. In addition, food subsidies are also provided by the state. The share of input subsidies (on fertilisers, power and irrigation) in the GDP rose sharply in the 1980s, and generally declined from the early 1990s. In the early 1990s, roughly 1.6 to 1.9 per cent of the GDP was spent on input subsidies, which fell to roughly 1.3 to 1.4 per cent between 2003 and 2006 (Ramakumar, 2009). On the other hand, the share of input subsidies in the agricultural GDP shows a general rise from the mid-1990s. In 1993-94, the share of input subsidies in the agricultural GDP was 6.7 per cent, while the corresponding figure for 2005-06 was 9.4 per cent. However, almost all of the increase in the share of total input subsidy to agricultural GDP after the mid-1990s appears to have originated from an increase in electricity subsidies. In fact, the shares of fertiliser subsidy and irrigation subsidy in agricultural GDP fell after the late-1990s and particularly in the 2000s.

The general, opinion is that subsidies are a problem, a view reflected in the views even of India's Planning Commission mentioned about it. However, the general argument given in favour of input subsidies is that it is necessary to encourage the use of particular inputs for production for various benefits.

Gulati and Sharma (1995) show that the input subsidy in India increased from 2.13 per cent of GDP in the triennium ending 1982-1983 to 2.73 per cent in the triennium ending 1992-1993. It is argued that the benefits of these subsidies have accrued to only certain classes of farmers in some regions cultivating irrigated crops. Furthermore, highly subsidized prices of inputs such as irrigation water and electricity for pump sets have encouraged cultivation of water-intensive crops, over-use of water, ground water depletion/salinity and water logging in many areas. Subsidy for nitrogen fertilizer on the other hand has resulted in nitrogen, phosphorus, potassium imbalance and acted as a disincentive for use of the environment friendly organic manure. As a result, the linkage between food crops and non-food crops, which include fodder, has been reduced. These adverse consequences are a drain on the fiscal burden of central and State Governments. Thus, if not properly monitored, input subsidies can be counterproductive and, in this context, protection to lower costs of production should be done selectively in the course of liberalization.

Ramkumar (2009) asks whether it would be desirable to withdraw input subsidies and if so at what cost and to whom. According to him, the fall in share of input subsidies in GDP after 1991 was the result of a conscious official policy that aimed at restricting subsidy provision, particularly for fertilisers (*cf.* Gulati and Narayanan, 2003). The argument in favour of reducing subsidies is put forward based on three reasons: first, subsidies constitute a substantial burden on the finances of the government; secondly, subsidies crowd out public investment by diverting resources; and thirdly, the prices of inputs do not reflect their scarcity value and hence these inputs are prone to overuse resulting in environmental degradation and fall in soil quality (Gulati and Sharma, 1995; Mahendra Dev, 1997; Vaidyanathan, 2000; Gulati and Narayanan, 2003; Gulati and Mullen, 2003). Ramkumar raises a valid point by noting that the "fiscal" argument for reducing subsidy is based on the *existing* levels of revenue-expenditure relationship of the government. The argument of fiscal stress may cease to hold once this relationship changes; that is, if the revenues of the government rise. In India, public provision of social and economic services has always been constrained by poor revenue collections. For instance, tax-GDP ratios in India are one of the lowest in the world.

The rationale for the provision of input subsidies has historically been to provide farmers with remunerative as well as stable prices so as to enable them to adopt new technologies and raise yields. Also, subsidies help to compensate for imperfections in the capital market and the risks associated with the adoption of new and high-cost technologies.

There is by now wide agreement that input subsidies have significantly aided the process of adoption of new technologies in the post-green revolution period. Sen (1992) argued that the agricultural growth of the 1980s was primarily due to a more intensive use of fertilisers and pesticides; there was a perfect negative correlation between yield of foodgrains and prices of fertilisers relative to foodgrains. Also, higher yields due to faster diffusion of technologies could restrain the growth of food prices to the extent that government and industry could keep their price indexed wage bills low (Acharya, 2000). There is also evidence that marginal and small farmers benefited significantly from input subsidies. Estimates in (Acharya and Jogi, 2004) show that 36.4 per cent of the total input subsidies were availed by marginal and small farmers, while their corresponding share in the ownership of operated area was 36 per cent. Acharya and Jogi also estimate that input subsidies were used mainly in the production of food crops, particularly rice and wheat. Given these benefits of subsidies, would it be desirable to withdraw subsidies, even if we assume that all of the reduced amount would be invested in agriculture? It appears certain that there would be significant short-term and adverse effects of withdrawal of subsidies on the levels of profitability in agriculture.

Climate Change and Food Security

Climate change is a reality in South Asia is a reality now. It is predicted to adversely affect millions across south Asia. In the Indian context, the projected impacts are an increase in average temperatures by 2-4 degrees Celsius, marginal changes in rainfall during monsoon months but large changes during the non-monsoonal months. The number of rainy days during the monsoons is projected to decrease by more than 15 days while the rainfall intensity is expected to rise by as much as 1-4 mm/day. Cyclonic storms are likely to increase in frequency and intensity (Mitra *et al.*, 2008).

Climate change potentially increases the vulnerability of the poor and those dependent on natural resources for their livelihoods. It leads to less secure livelihoods due to depleted social, financial, physical, natural resources and human assets; increasing health risks due to diseases like malaria, dengue, cholera, dysentery, malnutrition and exposure and constrained economic opportunities due to short and long term impacts of droughts and floods and extreme events. Women and children are the worst affected.

Technology Options for Achieving Food Security in South Asia

Technology has played a key role in changing agricultural production and thereby ensuring the food security. Various disciplines of agriculture, agricultural engineering, and agricultural chemistry have played a role in enhancing the agricultural production. Agricultural engineering has provided a wide array of agricultural equipments for tilling the land, pumping irrigation water from surface and subsurface region. Agricultural chemistry has played a significant role in the determination of fertility losses in the soil and the development of chemicals that can supplement soil fertility for a desired level of crop production. Various techniques and chemicals have also been developed for preventing the cropping environment from ecological damage.

In augmenting food supplies, genetic improvement through plant breeding has also played a great role. In India, a beginning has been made, and recently the Genetic Engineering Approval Committee (GEAC) of the Ministry of Environment has granted approval for BT cotton for commercial production. In the area of livestock production, technology has played a great role in boosting milk as well as egg and meat production. In India a kind of revolution has taken place. The story of white revolution, bringing India to the status of number one milk producer in the world, is well known. Thus technology and its effective applications are the answers for meeting the challenges of food security.

Conclusion

With complex regional, national, micro-macro, social and gender manifestations, the fact remains that the overall food security situation in South Asia is rather grim. With around 40 per cent of the population below the poverty line, a business as usual approach will not be sufficient to meet the emerging challenges of production, distribution and indeed governance.

There is an intense debate on whether farmers benefit more from commercial farming with modern technology or traditional farming by subsistence farming. Given the falling average size of holdings in South Asia, and that the vast majority of the hundred million holdings are small and marginal; solutions to enhance productivity on these are not going to be easy.

Recent farmers suicides in various Indian States show relying on export markets can bring disaster. In terms of distribution too the system of public distribution of foodgrains by the government has not achieved the desired results and has many distortions. A major gap in the research relates to the urban poor and the food security in the urban areas. While the nature and trends of urbanisation across South Asia is changing, with the phenomena of a high growth of peri-urbanism, unfortunately the research is yet to catch up with the trend.

REFERENCES

Acharya, S. S. and R. L. Jogi (2004). "Farm Input Subsidies in Indian Agriculture", Working Paper 140, Institute of Development Studies, Jaipur.

Acharya, M. P. M.; B. Acharya and S. Sharma (1999). *Women in Nepal: Country briefing paper.* Asian Development Bank. Manila.

Acharya, S. S. (2000). "Subsidies in Indian Agriculture and Their Beneficiaries", *Agricultural Situation in India*, 47 (5). pp. 251-60.

Bardhan, Pranab (1973). "Size, Productivity and Returns to Scale". *Journal of Political Economy* 81, No 8, November/December: 1370-1386.

Barret, C. B. (1996). "On Price Risk and the Inverse Farm Size-Productivity Relationship," *Journal of Development Economics*, 51, 193-215.

Beck, T. (1994). "Common Property Resource Access by the Poor and Class Conflict in West Bengal", *Economic and Political Weekly*, 29:4(1994): 187-196.

Bhalla, S. S. (1988). "Does land quality matter? Theory and measurement," *Journal of Development Economics*, 29, 45-62.

Bharadwaj, Krishna (1974). *Production Conditions in Indian Agriculture*, Cambridge: Cambridge University Press.

Chand, Ramesh (2004). "Impact of Trade Liberalization and related Reforms on India's Agricultural Sector, Rural Food Security, Income and Poverty", Institute of Economic Growth, New Delhi.

Chattopadhyay, M. and Ashok Rudra (1984). "Size Productivity Revisited — Inverse Relationship Science or Imagination", *Economic and Political Weekly* 19, Nos 52 and 53, December: A189-A 198.

Chayanov, A. V. (1926). "The theory of Peasant Economy", In Thorner, Daniel, H. Basile Robert Kerblay, E.F. Smith (eds), (1966). *The Theory of the Peasant Economy,* American Economic Association, Irwin Homewood, IL.

Choudhary, N. and D. Parthasarathy (2007). "Gender, work and household food security". *Economic and Political Weekly*, 42(6). 523-531.

FAO. (2002). FAOSTAT accessed at http://www.faostat.fao.org./default.htm; cited in Ramachandran, 2008.

FAO. (2000). *Gender and IPM in Nepal. FAO Programme for Community IPM in Asia.* Bangkok.

Gulati, Ashok and A. N. Sharma (1995). "Subsidy syndrome in Indian agriculture", *Economic and Political Weekly*, September 30.

Gulati, Ashok and Kathleen Mullen (2003). "Responding to Policy Reform: Indian Agriculture in the 1990s and After", Working Paper No. 189, Stanford Center for International Development, Stanford.

Krishnaraj, Maithreyi (2005). Food Security: How and For Whom? *Economic and Political Weekly*, June 18; pp. 2508-2512.

M. S. Swaminathan Research Foundation 1999.

Mahendra Dev, S. (1997). "Subsidies and Investments in Indian Agriculture: Issues and Perspectives", Rajiv Gandhi Institute for Contemporary Studies, New Delhi.

Mahesh, R. (2000). 'Farm Size Productivity Relationship: Some Evidence from Kerala' Working Paper No. 2; Kerala Institute for Environment and Development (KIED), Thiruvananthapuram.

Mitra, Amit and Nitya Rao (2009). "Displacing Gender from Development: A View from Santal Parganas, Jharkhand" in Mehta (Ed.) 2009.

Mitra, Amit (2008a): "Gender Caste and Growth Assessment – West Bengal Sub-National Study" (mimeo).

Mitra, Amit, Shashikant Chopde, Amit Kumar, Shiraz Wajih (2008). "Climate Change Adaptation Activities in India", Gorakhpur Environmental Action Group, Gorakhpur, Uttar Pradesh.

Radhakrishna (2005). "Food and Nutrition Security for the Poor in India', *Economic and Political Weekly*, April 30.

Rai, Parshuram (2005). "Political Economy of Hunger in *Adivasi* Areas" (mimeo), Centre for Environment and Food Security, New Delhi.

Ramachandran, Nira (2008). *Underfed, Underpaid and Overlooked Women: The Key to Food Security in South Asia*, International Institute for Development and Environment, Gate Keeper Series 8, February.

Ramkumar (2009). "Indian Agriculture under Economic Reforms" (mimeo). Tata Institute of Social Sciences, Mumbai.

Rao, Nitya (2005). "Gender Equality, Land Rights and Household Food Security—Discussion of Rice Farming Systems", *Economic and Political Weekly*, June 18, pp. 2513-2521.

Rao, Nitya (2006). "Women's Rights to Land: Gender Relations in Tenure – A Scoping Study in the Indian Context" *unpublished mimeo,* IDRC.

Rudra, Ashok (1968). "Farm Size and Yield Per Acre", *Economic and Political Weekly* 3, Special Number, July.

Saxena and Rachel Slater (2003). *Food Security and the Millennium Development Goal on Hunger in Asia*, Overseas Development Institute, London.

Sen, A. K. (1962). "An Aspect of Indian Agriculture" *The Economic Weekly*, Annual No. February, 243-266.

Sen, A. K. (1964): "Size of Holdings and Productivity", *The Economic Weekly*, Annual Number, February

Sen, Abhijit (1992). "Economic Liberalisation and Agriculture in India", *Social Scientist*, 20 (11). November, pp. 4-19, No. 21, May 22-28.

Srinivasan, T. N. (1972). "Farm Size and Productivity: Implications of Choice under Uncertainty," *The Indian Journal of Statistics*, 34, 409-420.2

Thapa, Sridhar (2007). "The Relationship between Farm Size and Productivity: Empirical Evidence from the Nepalese Mid-hills", CIFREM, Faculty of Economics, University of Trento, DECOS, University of Tuscia, Italy. Munich Personal RePEc Archive (MPRA) Paper No. 7638 posted March 11, 2008, Available online http:// mpra.ub.uni-muenchen.de/7638.

Thorat, Sukhdeo and Joel Lee (2005). "Caste Discrimination and Food Security Programmes," *Economic and Political Weekly*, September 24, 4198-4201.

Thorner, Daniel; H. Basile; Robert Kerblay and E. F. Smith (Eds.). (1966). *The Theory of the Peasant Economy,* American Economic Association, Irwin Homewood, IL.

Toufique, K. A. (1999). "The Relationship between Farm Size and Productivity in Bangladesh Agriculture: The Role of Transaction Cost in Rural Labour Market" Bangladesh Institute of Development Studies, pp. 55-71.

Vaidyanathan, A. (2000). "India's Agricultural Development Policy", *Economic and Political Weekly,* 13 May, pp. 1735-1741.

World Bank (2001). *Engendering Development through Gender Equality in Rights, Rresources and Voice,* New York, Oxford University Press.

World Development Report (2008). *Agriculture for Development,* The World Bank, Washington DC. 60.

Section V
Conclusion

15

Epilogue: Alternative Ways for the Small Farm

The modern history of Indian agriculture is the history of adopting modern scientific and technological practices emanating from a centralized knowledge system, which subsequently lead to centralized management of input supply, agricultural products and their distribution. A subsidy regime and excessive government interventions in input and product markets has been a key to establishing the highly expensive and ecologically disastrous agricultural production regime in place. The regime has created areas of abundance and scarcity. The food security bill is an outcome of these historical processes demanding a universal entitlement to all citizens of the country and for a centralized state to be duty bound to do this. Given the challenge to the environment and resources that the technological praxis and the climate change has posed, this symposium argued that there is an emergent need to consider a U-turn in agricultural policies to adopt a crop mix that is suitable to the environment and is near to the indigenous knowledge systems, ensuring sustainable food sovereignty. But this requires decentralized thinking and action, for which neither the governments nor the civil society advocating for food security are prepared. We present here excerpts from two presentations in the symposium, first by Vandana Shiva and the other by R. P. Singh.

I
Excerpt based on a presentation by Vandana Shiva

India has become the capital of farmer's suicide, of hungry people, of wasted and stunted children, of modern food related diseases of adultration in food and food products. No action is initiated against poisonous additives in industrial food and drink under the Prevention of Food Adulteration Act. Street food is also becoming equally poisonous. Such poisonous food is being promoted while the fresh food prepared by mothers at home is despised by school children. Sub-standard food is termed as safe for health and fit for marketing. Our diversified food is nowhere in focus. Our healthy mustard oil is out but imported zero duty soya oil is in. Our *dal* (pulse) is out from our farms but *I-dal,* reconstituted soya, is welcomed by our westernized rich women.

With the help of World Trade Organization (WTO) driven initiatives Multi-National Corporations (MNCs) like Cargill, Monsanto, are penetrating deep in the country's agri-business with mighty lobby and are dictating the terms of agricultural production and trade. Prices of seeds are determined by these MNCs under the IPRs regime. The local nutritious diversified products like *Basmati, Neem*, Melon, Mango, *Kinnoo,* Tomato, etc., are being genetically modified and patented by these Corporations. The whole system is now locked leading to steep increase in cost of seeds. Around 95 per cent cotton seeds sold in the country are now locked with a price escalation by 8,000 per cent during past 10 years. Imposing IPRs on seeds is equal to '*lagan*' (tax) on life. This is leading to farmers suicides mainly due to very high input costs and high costs of seeds. The Genetically Modified Organisms (GMO) seeds are now sold in India in the name of technology. All these GMO varieties removed of ageing genes, however, are like stones, deprived of softness and nutrition contents. Net result is, enormous varieties of new weeds in the country. The entire native food web is being turned into market commodity chain.

Small farmers produce more nutritious and ecological variety food. The food variety contains around 20 to 25 diverse items in 1.2 billion Indians' *Thali* per meal, while Italy in the west which is listed in United Nations Educational, Scientific and Cultural Organization (UNESCO) heritage list for diversified food, ends up with only 4 to 5 items of food. The bio-diverse organic *versus* mono-culture chemicals genetically engineered food products are leading to a disaster. The domestic variety of manure is disappearing, while chemical fertilizers are imported from Cargill with a net result of less nutritious commodity production in all respects including protein, calcium, iron, fiber. Such is the nature of the western agriculture. The new *mantra* now being propagated is that 'the small farmer is non-viable'. This is nothing but to initially kill him and subsequently accuse him of committing suicide. We allowed MNCs for retail trade through super markets from 2005 onwards. As a result farmers' incomes halved and MNCs profits doubled and Rs. one trillion is the estimated loss per year now. The chemicalized and industrialized agriculture is leading to dangerous consequences in India. Our small farmer may become food sovereign once again, if we honestly initiate our own research for knowledge sovereignty. We need to build our food sovereignty; otherwise more farmers will opt for suicides in future. *Navdanya,* is trying to develop the bio-diverse indigenous model for agriculture.

II
Excerpt from a presentation by R.P. Singh

Food supply is increasingly controlled by giant multinational corporations and global large agribusiness corporate houses which significantly affect the global agricultural economy. "Free trade" agreements are turning food into a commodity for profit. This has disastrous effects on farmers, food security, and the environment. Even in the developed countries various trade agreements and dominance of multinationals have adversely affected the small land holders as millions of small farmers in Canada, USA and Mexico have lost their land while agribusiness corporations reaped huge profits. Globalization of agriculture would put transnational corporations to control the entire food supply and market forces, rather than national policies set by democratically elected representatives. Under this scenario, each country would be forced to only produce a few export commodities, wiping out local food production, small family farms, and greatly compromising global food security. As a result, the human right to food would be dependent on multinational corporations and markets, increasing the risk of hunger and famine worldwide.

The multinationals and supermarkets, worldwide own a growing part of the market and control the trade in fresh vegetables and fruit. In the Netherlands, more than 80 per cent of all fresh fruit and vegetables are sold at supermarkets, of which the 5 largest supermarkets own more than 50 per cent of the market. Similarly, a mere 2 per cent of farms in the United States control more than half of American agricultural sales. Over 73 per cent of the nation's farms share less than 7 per cent of the market value of agricultural products, while 7.2 per cent of farms receive 72 per cent of the market value of products sold. Around 82 per cent of U.S. corn exports are controlled by three agribusiness firms — Cargill, Archer Daniels Midland (ADM), and Zen Noh. Family farmers' incomes have plummeted during the first 7 years of North American Free Trade Agreement (NAFTA), but ADM's profits went from $110 million to $301 million, and ConAgra's grew from $143 million to $413 million.

The increasing concentration of financially strong chains of foreign multinational supermarkets has an effect deep into the supply chain of agriculture as they dictate who produces what, where and for what prices. The small farmers find it difficult to meet their requirements of best quality products at low prices and that too in large quantities throughout the whole year. Market Access goal of agribusiness interests in 'free trade' negotiations is mainly to eliminate quotas and tariffs so that they can gain access to foreign markets for their products. Since most of the agribusiness corporations are from rich countries in the global North, market access will generally benefit them at the expense of small farmers in poor countries.

The government instituted subsidies to farmers, actually benefit multinational corporations by keeping prices low while taxpayers foot the bill. Benefits of subsidies to corporate agribusinesses allow them to buy crops from small farmers at a price far below the cost of production and allow Multinationals to sell at market rates. Though, subsidies provide some relief to limited number of struggling family farmers, they are not enough to make up for low commodity prices but contribute to the erosion of small farms. In fact, the prices that farmers receive today for their crops are likely to be well below the cost of production, causing millions of farmers to lose their land and go further into debt. Moreover, the policies driven by corporate globalization are pushing farmers off the land, and peasants out of agriculture. This is not a natural evolutionary process but a violent and imposed process leading to increased farmers' suicides.

Summing Up

The small farmers in India today are confronting numerous problems: Imperfection in input and product markets is affecting profitability; access to institutional credit, technology, energy and water supply is inadequate. The country is facing farmers' suicides and a large number are affected by food related diseases. IPRs are imposed on seeds resulting in several times increase in seed prices, one of the major factors leading to farmers suicides mainly due to very high input costs and high costs of seeds. The GMO seeds are now sold in India in the name of technology. All these GMO varieties are deprived of softness and nutrition contents. The entire native food web is being turned into market commodity chain. Food supply today is increasingly falling under the control of giant multinational corporations which significantly affects the global agricultural economy, using "free trade" agreements to turn food into a commodity for profit. The increasing concentration of financially strong chains of foreign multinational supermarkets has an effect deep into the supply chain of agriculture as they dictate who produces what, where and for what prices. The small farmers find it difficult to meet their requirements of best quality products at low prices and that too in large quantities throughout the whole year. The policies driven by corporate globalization are pushing farmers off the land, and peasants out of agriculture. Our small farmers may become food sovereign once again, if we honestly initiate our own research for knowledge sovereignty.

Small farmers are intensifying biodiversity through conservation and rejuvenation resulting into higher biological productivity and stability and sustainability of agriculture. Unfortunately, the small farmers today face several constraints physical, socioeconomic, technological, infrastructural, and institutional. Region-specific approaches for agricultural development may be viable along with their integration with rural development policies. Diversification of cropping system is required with focus on those traditional coarse crops which have more nutritional values, also require low inputs and water as well as withstand drought/ moisture stress along with emphasis on livestock based farming system. Small farm producers, farm workers, consumers, and their organizations may be involved and invested with real decision-making power in trade negotiations and issues relating to market chains that affect their futures. Seeds, plants, animals, and their components may be exempted from patenting. Agricultural policy must preserve indigenous knowledge of farmers related to agriculture methods, use of seeds and plants and utilize their cultural knowledge and collective wisdom in the use of resources.

Small farmers preserve bio-diversity. They could conserve, recycle soil nutrition, rejuvenate eco systems and ensure sustainability of agriculture. The myth of chemical-industrial-large sized agriculture needs to be broken with more research and if needed experimental research on the diverse crop of the tropics and their nutritional values.

Index

Adani group, 31
Agrarian Reforms Committee, 160
Agrarian reforms, 230
Agrarian Society, 222
Agribusiness system, 72
Agricultural commodities, 3, 10, 39, 78–79, 103, 144, 244
Agricultural credit, availability of, 175
Agricultural development, 6, 39, 156, 178, 180, 204, 230, 235, 263
Agricultural food system, 100–101
Agricultural Market Committee, 81
Agricultural marketing system, 22
Agricultural policies, 7, 11, 103–105, 193, 225, 230, 237, 259, 263
Agricultural Prices Commission, 181
Agricultural Produce Market Committee (APMC), 8
Agricultural prosperity, 204
Agricultural reforms, 20
Agricultural security, 7
Agricultural society, 230
Agricultural technology, 92, 113, 166
Agriculture
  commercialization of, 123, 130
  contribution of, 173, 201
  declining contribution of, 13, 77
  development of, 13, 173–176, 180, 182–183, 198
  diversification of, 11, 112
  globalization model of, 233, 235
  growth of, 12, 113, 142, 151, 198
  industrialized, 261
  supply chain of, 262–263
  sustainability of, 6, 98, 263
Agro Services Centres, 180
Agro-climatic zones, 13
Agro-economic Research Centres (AERCs), 22
Agro-ecosystems, 201
Agro-forestry elements, 203
Agropastoralism, 124
Animal husbandry, 120–, 127, 168, 201
APMC Acts (APMCA), 19, 22, 28
Archer Daniels Midland (ADM), 100, 261
Artificial world simulation model, 49
Assam Agricultural University (AAU), 22, 24–25
Awareness campaigns, 181
Awareness programmes, 79

Backward groups, 64
Balance diet, basket of, 194
Bargaining power, 5
Bhattacharya Committee, 180
Biochemical technology, 138
Biodiversity, 6, 14, 52, 59, 98, 189, 193–196, 199, 202, 215, 234–235

farming system, 210
indigenous model, 261
Bio-fuel programmes, 234
Biological system, 188
Biotechnology, 245
Bombay Cotton Trade Association Ltd, 77
Broad-based schemes, 198
Bundelkhand, 170

Capital formation, 179
Capitalism, 232
Capital-scarce country, 6, 93
Cash crops, 47, 78, 118, 166, 180, 237
Caste category, 124
Caste conflict, 230
Casual employment, availability of, 130
Catalytic factors, 176
Cell phones, 138
Central Statistical Organization (CSO), 75
Centralized knowledge system, 259
Chemical-industrial-large, 263
Chicago Board of Trade (CBOT), 76
Chilli, 79–84
cultivation, 80
dried, 80
exchange-based Trading of, 81
export, 80
exporter of, 80, 82
farmers, 85
futures contracts, 81
market, 79, 82
oleoresin, 80
pickled, 80
powder, 80
price of, 81, 84
purchasers, 81
receipt, 81
trade, 80, 82–84
traders procure, 81
value chain, 83
Chi-square tests, 62, 64–65, 67
Civil society, 229
Climate Change Compliant Crop (CCCC), 14, 202, 210, 217
Climate change, 16, 201–203, 206, 210, 252–253, 259
Climate friendly food system, 202
Climate system, 202
Climatic factors, 138, 159
CMI model, 22
Coca-Cola Company, 224
Cold Chain System, 198
Commission on Agricultural Costs and Prices (CACP), 209–210
Compound Annual Growth Rate (CAGR), 158
Constitution of India, 45, 242
Construction sector, 125
Consumerism, 237
Contemporary society, 194
Contract farming, 7–8, 20, 39, 43, 97, 237
Cooperative banks, 98, 182
Cooperative/government agencies/ commercial banks, 168
Corporate agents, 22
Corporate agribusinesses, 101, 262
Corporate agriculture, 233
Corporate buyers, 28
Corporate farming, 15, 238–239
Corporate intermediation, 8, 35
Corporate Marketing Intermediation (CMI), 20, 26, 32, 34–35
Corporate marketing, 20, 34
Corporate social responsibility, 217
Corporations, 101, 229, 234, 260
agribusiness, 100, 261–262
foreign, 234
multinational, 100–101, 105, 261–263
powerful, 236
transnational, 100–101, 225, 237, 261
Credit institutions, 133, 176
Cretinism, 191

Critical factors, 143
Crop
  diversification, 67, 180, 232
  production, 39, 176, 189, 198, 253
  productivity, 92
  rotation, 67, 72
Cropping patterns, 38, 47, 175
Cropping system, 11, 103, 263
  diversification of, 11, 103, 263
Cultivable
  cost of, 4, 33, 46, 161, 163, 165–166, 169
  extension of, 11, 112
  land, 94, 118, 132
  of fruits, 180
Dalit society, 238
Decentralized programmes, 176
Deepak Fertilizers and Petro-Chemicals Limited (DFPCL), 23–24, 30–31
Democratic society, 137
Demonstration centres, 180
Department of Foods and Nutrition, 213
Depeasantisation process, 4, 8, 15, 238–239
Development
  balanced, 12
  bio-fuel, 234
  constituted, 223
  dairy, 114
  entrepreneurship, 150
  evolutionary, 51
  inclusive, 201, 222
  indicators, 117
  infrastructural, 222
  integrated, 149, 205
  of agriculture sector, 250
  patterns, 114–115
  planned, 146, 227
  product, 209
  specificity, 116
  technological, 42, 96
  trajectory, 120
Developmental programmes, 149–150
Developmentalism, 235
Diet revolution, 245
Direct marketing, 8, 20, 26, 28–30, 32, 34–35
Disasters, 254, 260
Discrete mathematics model, 8
Disparities, 236
Distribution systems, 245
District Industry Centre, 147
Domestic life support system, 46
Domestic prices of cereals, 250
Drought prone areas, 208
Droughts, 175, 188, 205, 253

Eastern and Central Africa (ECA), 206
Economic activities centre, 174
Economic development, 44, 117, 118, 229
Economic empowerment, 238
Economic environment, 249
Economic liberalization, 39, 228–229
Economic mechanism, 169
Economic reforms, 15, 89, 222, 228, 232, 236
Ecosystems
  disruption, 235
  fragile, 201
  resilient farm, 203
Education
  low levels of, 129
  vocational, 132
Egalitarian society, 222
Electronic Price Display Boards, 84
Elite-oriented model, 235
Empirical tests, 40
Employment Generation Programmes (EGPs), 150–151
Employment
  alternative sources of, 95, 175
  casual type of, 126–127
  intensity of, 124
  multiplicity of, 121–122
  off-farm, 39

Enron Company, 224
Entrepreneurial activities, 133
Entrepreneurial dynamism, 230
Environmental degradation, 9, 53, 188, 251
Environmental economics, conventional methods, 44
Essential Commodities Act, 235
Evolution and Reforms of Markets, 21
Evolutionary economics rules, 41
Export markets, 30, 254
Extension programmes, 226
Extension systems, 226

Factor productivity, 144
Family labour, 133, 159, 164, 167
Farm families, upward movement of small, 6, 95
Farm incomes, expansion of, 170
Farm operations, 181–182
Farm sector, 12, 129, 142, 174, 180–182
Farm technology, 70
Farm women, 189
Farmer's
  affluent categories of, 225, 238
  associations, 198
  Exchange System, 197
  group, 5, 29
  Interest Groups, 11
  off-farm income of, 140, 142, 145, 149
  organizations, 6
  viability of small and marginal, 144, 150
Farming of animals, 139–140, 161
Farming system, sustainability of, 6, 90
Farm-related factors, 72
Female empowerment, 204, 216
Fertilisers, 236, 250–252
Fertility, 168, 212
Field survey, 167
Finger millet
  feasibility of, 14
  development of, 15, 215, 217
  Inclusion of, 197
  composition of, 211
Fiscal policy, 143
Fisheries post-harvest operations, 170
Fishing, 189, 201
Five Year Plans, 150
Floods, 4, 178, 188, 253
Food Adulteration Act, 260
Food and Agriculture Organization (FAO), 188, 190, 242–244
Food and Civil Supply and Consumer Welfare Department, 176
Food availability, 188, 193
Food gathering, 189
Food insecurity, 14, 16, 77, 122, 189, 196, 235, 237, 244
Food Policy, 13
Food scarcity, 244
Food Security Act, 243
Food security, 7, 12, 15–16, 46, 90–91, 188–, 210, 242–244, 253
  assurance of, 13, 173
Food sovereignty model, 223
Food sovereignty, 8, 100, 261
Foodgrains sector, 248
Forest Department, 182
Forward Markets Commission, 84
Fragmentation of farms, 105
Free trade agreements, 100, 261, 263
Fuel wood, 189

Genetic Engineering Approval Committee (GEAC), 253
Genetically Modified Organisms (GMOs), 105, 260
Global assessment report, 204
Global Conference on Agricultural Research for Development (GCARD), 244
Global Hunger Index, 242
Globalization era, 233, 236
Globalization of agriculture, 100–101, 261

Gokhale Institute of Politics and Economic, 22
Gramya and Cooperative Banks, 182
Graph theoretic model, 49
Graph theory, 41, 57–58
Green revolution crops, 201, 216
  expansion of, 201, 216
Green revolution, 5, 13, 38, 46–47, 53, 89, 105, 143, 159, 187, 204, 216, 230–231
Gross Value Added (GVA), 148
Group marketing, 29
Group rationality, 42
Growers' Representative Group (GRG), 25, 29
Growth rate index, 214

Harmony, 140, 237
Harvesting seasons, 124–125
Healthcare, 243–244
Heart disorder, 192
High Yielding Variety (HYVs), 46–47, 122, 165, 231
High-value vegetables, 9
Hill agriculture, 122
Hill areas, 122, 124, 130
Hill economies, 12, 113–114, 118
Himachal Pradesh University (HPU), 22
Hired labours, 133, 167
Household industries, 12, 113
Household survey, 131
Human Development Index, 242
Human resource development, 132
Hunger, risk of, 100–101, 261
Hydroelectric project, 114
Hydro-power, 115
Hypo-thyroidism, 191

Income Quintile Groups, 129
Indebtedness, Incidence of, 47, 110
Indian Council of Social Science Research, 7
Indicators, 32, 62
Industrial activities, 117, 119
Industrial development, 116, 120
Industrial economy, 222
Industrial estates, 147
Industrial farms, 6
Industrial food, 260
Industrial Policy, 117
Industrial potential survey, 147
Industrialists, 148
Industrialization, 46, 119, 230, 232
  rural, 148
Industries
  handlooms, 116
  heterogeneous, 149
  local brewing, 213
  non-farm/small scale, 169
  rural, 147
  small, 146
  traditional beer brewing, 208
Inequalities, 15, 229, 231, 238
Infant Morality Rate (IMR), 115
Information and Communication Technologies (ICTs), 70
Information technology, 245
Innovative practices, adoption of, 64, 69–70
Input-output model, 48, 50
Institute for Food and Development Policy, 93
Instituted subsidies, 101, 262
Institutional constraints, 9, 95–96
Institutional credit, 5, 72, 97–98, 155, 175, 262
Institutional development, 79
Institutional environment, –, 43
Institutional framework, 231
Institutional mechanism, 131
Institutional order, 229
Institutions
  academic, 215
  agricultural, 167
  contemporary, 20
  economic, 42
  formal market, 68

global, 233
human, 43
public sector marketing, 26
rural, 151
trans-national, 15, 224
Integrated pest management (IPM), 67, 72
Intensive Agriculture District Programme (IADP), 46
Inter-Ministerial Committee, 22
International Crops Research Institute for the Semi-Arid Tropics (ICRISAT), 208–209
International Crops Research Institute, 208
International Food Policy Research Institute (IFPRI), 7, 242–243
International Labour Organization, 228
International Monetary Fund, 15, 224
Intestinal parasites, 244
Intra-model comparison, 35
Iodine Deficiency Disorders (IDD), 191
Iodine deficiency disorders, 191
IPCC Report, 202
Irrigated crops, 251
Irrigation, 11, 39, 62, 70, 72, 95–96, 165, 168, 204, 232, 250

Jajmani system, 213
Jawaharlal Nehru Krishi Vishwavidyalaya (JNKV), 22
Jharkhand, 90, 97
Jhumba rice, 99
Joint farming system, 167

Karnataka Rajya Raitha Sangha (KRRS), 224
Khadi and Village Industries Corporation (KVIC), 146
Kisan Credit Cards (KCC), 98, 176
Kisan Vikas Kendras (KVKs), 84
Knowledge-based technology, 138
Knowledge-intensive technology, 138

Labour markets, 120, 122, 222
Land Acquisition Act, 227
Land capability, 113, 134
Land degradation, 38, 231
Land reform policies, 103–104
Land reforms, 46, 92, 156, 160, 236–238
agenda of, 236–237
Landless families, 125
Landless labourers, 236
Land-man ratio, 95
Land-use pattern, 231
Land-use sectors, 202
Leontief system notations, 57
Lewisian development framework, 120
Liberal agro export model, 223
Liberalization, 5, 223, 225–226, 228–229, 234–235, 237, 249, 251
concomitant, 224
external, 234
internal, 234
policies, 228
Livelihood generation programmes, 244
Livelihood, source of, 13, 174, 245
Livestock, 12, 103, 113, 137, 170, 175, 203–204, 215, 245, 263
rearing, 11, 112, 142
quality of, 11, 112
Local markets, 29, 133
Low Birth Weight (LBW), 190
Low cost technology, 72

Mahatma Gandhi National Rural Employment Guarantee Act (MGNREGA), 142, 182
Maize, 14, 99, 196–197, 203, 206, 212
Malaria, 253
Malnutrition, 14, 16, 189–190, 192, 211–213, 235, 244, 253
Manures, 161, 166–167, 260
Marginalized groups, 15, 224, 228–229

Market
  forces, 21, 100–101, 228, 261
  liberalization, 26
  reforms, 19, 32
Marketing federations, 10
Matrix-spanned graph, 57
Micro Finance Institutional, 176
Micro, Small and Medium Enterprises Development (MSMED), 146
Mid-day meal, 215
Migration, 39, 142, 182
Millennium Development Goals, 91
Minimum Support Prices (MSP), 77, 104, 155–156, 164, 169, 250
Minor millets, 203, 205–207, 216
Money lending landlords, 44
Moneylenders, 5
Monthly per capita income (MPCI), 121
Movements of landless, 223
MSMED Act, 146
Multilateral organizations, 150
Multi-National Corporations (MNCs), 224, 260
Multiplicity of activities, 122, 141

Nanotechnology, 245
National Agricultural Research System (NARS), 208
National Alliance of People's Movements (NAPM), 224
National Commodity and Derivatives Exchange (NCDEX), 78–79, 82
National Council of Applied Economic Research (NCAER), 93, 228
National Family Health Survey (NFHS), 190, 242
National Food Security Mission, 210
National Multi Commodity Exchange (NMCE), 78
National Rain-fed Areas Authority (NRAA), 205
National Research Institute (NRI), 208
National rural livelihood mission, 198
National Sample Survey (NSS), 38, 137–139, 148, 154
Nationalism, 232
Natural Calamities Relief Fund, 102
Natural ecosystem, 235
NCDEX Chilli Futures Contract, 81, 83
NCEUS Report, 166
Nehruvian ideology of industrialization of India, 230
Nehruvian socialism, 230
Neoclassical tax-transfer economics, 40
Neo-weberian theory, 226
Net sown area, 178–179, 210, 232
Net State Domestic Product (NSDP), 115
Network theory, 49
New Agricultural Policy (NAP), 9, 38–40, 43
Non-agricultural labourers, 176
Non-farm
  development agencies, 147
  income, 4, 140, 250
Non-Governmental Organisation (NGOs), 9, 183
North American Free Trade Agreement (NAFTA), 100, 262
Nutrition education, 193
Nutritional awareness programmes, 193
Nutritive food system, 217

Off-farm activities, 161, 169
Off-farm income, 139–142, 150–151
  increasing, 142
Open auction method, 81
Open market operation, 250
Operational efficiency, 180
Operational landholdings, 70, 157, 158, 160
Organic community, 233
Organic food bonus, 14, 217
Organic manure, 7

Own account enterprises (OAEs), 146

Paddy, 99, 155–156, 163–164, 166, 176, 187
Panchayati Raj Institutions (PRIs), 105, 150, 199, 216
Parsaul villages, 227
Participatory development, 150
Peasant nationalism, 232
People's Movements, 223–224, 239
Peri-urbanism, 254
Personal interview, 61
Phytosanitary technology, 4
Planning Commission, 94, 117, 201
Poisson Count Models, 71
Poisson Count Regression Model (PCRM), 62, 69
Policy
  agenda, 7
  alternatives, 50
  constraints, 238
  distortion, 143
  imperatives, 114, 173
  implications, 53, 72, 102, 130
  interventions, 44, 46
  outcomes, 40
  reservation, 143, 146
Political power, 45, 222, 231
Poor post-harvest system, 208
Post-green revolution period, 252
Post-harvest practices, adoption of, 68–70
Post-harvest technology, 72, 215
Post-reform markets, 19
Post-reform period, 179, 182, 232
Poverty
  ratio, 116
  reduction, 201, 214, 233, 245
  stricken mass, 45
  trap, 238
Pre-reform period, 232
Pressure groups, 43
Prevention of Food Adulteration Act, 260
Price discovery mechanism, 82
Price indexed wage bills, 252
Primary data, 8, 20, 35
Primitive marketing system, 28
Privatization era, 15, 224
Productivity, Low levels of, 188
Protein energy malnutrition (PEM), 190
Public announcement system, 81
Public distribution system, 14, 197, 217, 235, 254
Public expenditure, Efficiency of, 143
Public information system, 28

Questionnaire, 20

Rao, Hanumantha, 92
Rashtriya Krishi Vikas Yojana, 215
Rational formulation, 175
Rationalization, 144
Recycle soil nutrition, 263
Reform period, 250
Reform policies, 169
Regional disparities, 98, 233
Regional factors, 32
Regional Rural Banks (RRBs), 98, 176, 182
Regression model, 63
Regular marketing system, 28
Research methodology, 82
Resource endowments, 140, 147
Resource-based organic manufacturing, 146
Rich farmers, 210, 228
Royal Commission on Agriculture, 21
Rural communities, 234
Rural credit system, 176
Rural development, 150, 210, 228
Rural Economic Development Survey (REDS), 93
Rural Non-Farm (RNF), 12, 132–133, 143–149, 170
Rural population, 15, 91, 175, 234–235, 238

Rural poverty, 13, 32, 77, 116
Rythu Bazaar (RB), 24, 29

Sectoral policies, 143, 148
Sectors
agro-commodity, 77
allied, 145
commodity, 78–79
co-operative, 177
corporate, 26
dairy, 145
labour-intensive, 228
labour surplus, 182
non-agricultural, 140
predominant, 128, 248
tertiary, 118
Self-Help Groups (SHGs), 7, 29, 103, 234
Service sector, 104
Settled agriculture, 118, 188
Shifting cultivation, 118
Simple graph theory model, 48
Simplified agricultural system, 203
Situation Assessment of Farmers (SAF), 137, 148
Small farm model, 103
Social science research, 7
Social sector, 150
Social transformation, 229
Socio-economic factors, 62, 204
Software Package for Social Science (SPSS), 62
Soil
degradation, 26
degraded, 203
fertility, 99
food web, 168
less fertile, 205
management practices, 175
quality, 20, 46, 167, 251
saline, 210
sandy, 207
types, 207, 217
Special Economic Zone (SEZ), 102, 224
Spices, 14, 78–80, 82, 194–196
Staple food, 60, 208, 237
State Gross Domestic Product (SGDP), 173
State-centred developmentalism, 230
Storage facilities, 96, 104, 180–181
Structural adjustment programmes, 249
Subsidies, Benefits of, 101, 262
Suicides, 4, 7, 15, 47, 99, 102, 238, 261–262
Super Markets, 3, 5–8, 100–101, 245, 261
Surplus manpower, 11, 104
Survey of India, 93
Sustaining Himalayan Ecosystem, 203
Systematic analysis, 13

Technology demonstrations, 216
Technology generation, 149
Technology transfer, 179
Tenancy reforms, 5, 158
Tendulkar Committee, 94
Tourism, 12, 114–117, 119, 146–148, 151
cultural, 147
diversification, 116
horticulture-based, 147
industry, 115, 117
rural, 147
Trade liberalization, 20, 229, 232, 234, 249–250
impact of, 249
Trade unions, 224
Trans-National Corporations (TNC), 225, 237
Transportation, 77, 96
Turmeric, 77, 79, 194
Two-rupee rice schemes, 182

UK National Research Institute, 208
Under-developed seed system, 209
Underdevelopment, 119
Unemployment, intensity of, 129

United Nations Educational, Scientific and Cultural Organization (UNESCO), 260
United Progressive Alliance (UPA), 9
United States Department of Agriculture (USDA), 243
Universal Declaration of the Human Rights, 242
Unorganised Manufacturing, 148
Urbanization, 232
Uttarakhand, 99, 112, 114, 116–120, 124, 128–131, 192, 205
Uttarakhand hill economy, 127–128
Uttarakhand hill region, 119

Value chain, 7–8, 81–82
Vegetable
  chain, 70
  crops, 96, 104
  cultivation, 60, 64, 116
  dyes, 145
  farmers, 61, 64
  fresh, 100, 261
  growers, 9, 61–62, 64, 66, 70, 72
  leafy, 195
  marketing of, 9, 70
  minor, 195
  production, 60–61, 131
  sectors, 61
  supply chain, 62, 64, 69–70
  value chain, 61, 72
Village industries, 158
Vivekanand Hill Agriculture Research Institute, 210
Voluntary organizations, 149–150

Wage labour, 47, 182
Wage-based programmes, 150
Water resources, 38
Water revolution, 159
Weeding, 125, 209
Wheat and Sugarcane Crops, 155–156
White revolution, 253
Women
  groups, 198
  multiple roles, 189
  nutritional status, 193
  westernized rich, 260
World Health Organization, 192
World Trade Organization (WTO), 15, 104–105, 224, 233, 250, 260
Worldwatch Institute Report (WIR), 202–203

Zhengzhou Commodity Exchange (ZCE), 76